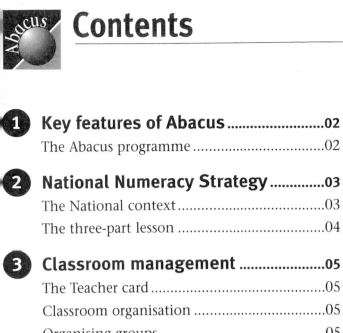

Contents

KU-788-993

1 Key features of Abacus

The Abacus programme

Abacus follows the detailed plan of teaching objectives outlined in the Framework for Teaching Mathematics. Furthermore, Abacus, from its inception, has been designed and written with precisely the same approach to the teaching of mathematics as that of the National Numeracy Strategy. Abacus is based on three principles:

Direct and interactive teaching is at the heart of the process of helping children learn mathematics. Children often do not 'discover' strategies; they have to be taught them.

Many mathematical skills and facts, particularly those which help children become fluent in mental calculations, need to be taught clearly, and then rehearsed regularly on a 'little-and-often' basis.

Materials need to support teachers in their teaching, and to help keep classroom management simple and effective. It is not necessary, nor is it desirable, for every classroom teacher to 'reinvent the wheel'. A clear structure of key objectives (with sufficient flexibility for important professional decisions to be left to each teacher in their own specific context) will minimise the hours spent planning and preparing, and maximise the teacher's effectiveness in the classroom.

As well as being based on the same philosophy, Abacus shares several assumptions with the National Numeracy Strategy.
- There will be a daily mathematics lesson.
- The lesson will have a three-part structure. Although the timing will be flexible, and will vary from lesson to lesson, class to class and school to school, the key elements will be common to all years and to all classroom contexts.
- Planning will be carried out on a weekly basis, with reference to the medium term planning grids in the Framework for Teaching Mathematics.
- In their direct teaching, with the whole class, in groups or with individuals and pairs, teachers will require simple practical resources. These will include large number lines, large number grids, medium number lines and small number grids, pupil sets of digit cards and place-value cards.

② The National Numeracy Strategy

The National Context

The National Numeracy Strategy follows the recommendations of the final report of the Numeracy Task Force (1998) that throughout the primary phase in education there should be:

- a daily mathematics lesson of 45 to 60 minutes
- a greatly increased proportion of whole class interactive direct teaching of mathematics
- a focus on teaching mental calculation strategies.

At the heart of the National Numeracy Strategy is the Framework for Teaching Mathematics. This document provides a structured sequence of teaching objectives for each year. Put simply, it outlines the mathematical content which must be taught to each school year group. Teachers are wholly aware of the necessity for teaching a structured sequence of skills to children, particularly in relation to developing effective mental calculation strategies (we are all familiar with the problem of teaching a topic, only to find that an essential pre-requisite skill has not been covered). The Framework provides just such a programme. From September 1999, all schools in England will be expected either to draw upon the Framework for Teaching Mathematics in planning their lessons, or to ensure that their teaching programme covers a similarly complete and structured sequence of teaching objectives, with a focus on the development of mental calculation strategies.

The three-part lesson

In the National Numeracy Strategy and in Abacus, the daily mathematics lesson has the following three-part structure.

Part of lesson	What it comprises	Content includes resources	Abacus
Oral/ mental starter	Usually number work Sometimes related directly to the main teaching activity and sometimes not.	Counting (in steps of 1000, 50, 0.1, 0.5 etc.) Strategies which have already been taught Number facts (+, −, ×, ÷)	*Mental Warm-up Activities* *Simmering Activities*
Main teaching activity	One of the following: Whole class introduction to the topic, using some paired work Group work Paired investigations Individual practice	Selected objectives from an appropriate year of the Framework for Teaching Mathematics	*Teacher Cards* *Activity Books* *Textbooks* *Photocopy Masters*
Plenary	Validation and report on the work of children with whom you haven't worked directly Rehearsal of main teaching objective in the light of any common misconceptions Explanation of the content to be covered next	Summary of key objectives Outline of objectives to be addressed next Reinforcement activity Possible homework activity	*Teacher Cards* *Simmering Activities* *Homework Books*

There is a variety of possible formats for the middle part of the lesson, depending on:

● where in the particular topic the lesson takes place (e.g. introducing a topic, continuing or extending a topic, assessing a topic, or revising a topic)
● the nature of the mathematical topic being taught
● the age and maturity of the children
● the nature of the classroom and school (e.g. mixed-age class, small rural school, large urban school, special school or unit).

The three-part structure is flexible, and is designed to accommodate teacher's individual teaching styles as well as differences in situation, organisation and content.

It is assumed (by the National Numeracy Strategy and by Abacus) that at the start of a topic there will be more whole class teaching, whereas in the middle or towards the end of a topic, there will be more directed teaching, focused on groups, pairs or individuals. However, over the topic as a whole, there will be a substantial proportion of direct teaching of the whole class (or of the whole year group in mixed-age classes).

❸ Classroom management

The Teacher Card

For each unit of work the Teacher Card identifies for you:
- the direct teaching (introductory and follow-up) to the whole class or group
- whole class activities and small group differentiated activities
- appropriate Textbook pages, Photocopy Masters and Mental Warm-up Activities
- ideas for Plenaries.

This enables you to plan effectively for the whole class (large or small), covering a wide range of ability.

Children will experience a range of tasks within each unit, the actual composition determined by you following the guidance on the Teacher Card.

Classroom organisation

Abacus allows for flexibility in the organisation of the children. On the assumption that the teaching takes place with the whole class, this can be followed using a variety of different organisational structures. Examples include:
- the whole class working together on the same activity, possibly in pairs
- half of the class working on one Activity, whilst the other half work on another Activity
- a 'carousel' of Activities, with the children in groups, moving from one activity to another
- half of the children working in groups on Activities, whilst the other half are working on material in the Textbooks or Photocopy Masters.

Organising groups

- You will decide on the best organisation of groups to suit your needs at any time. There should be no more than three levels of work in a single-age class, and no more than four levels in a mixed-age class.
- The composition of the groups will vary depending on the unit of work being studied.
- You can plan within your group structure, how, during the unit, you can work in a focused way with any one group.
- You will need to judge how much time to allow for different children to complete any task. The carousel nature of the tasks provides versatility in management.
- Other adults can be used to assist with the activities.
- Some children in the class may be used occasionally in a peer-tutoring role within the activity carousel.

4 The Abacus materials

Mental Warm-up Activities

The Mental Warm-up Activities Book provides a comprehensive scheme of work for developing mental mathematics strategies. Each day you will select an appropriate activity to take place before the main part of the lesson. The activities are designed to rehearse and sharpen key mental mathematics skills, including counting, comparing and ordering numbers. They allow you to get off to a clear, crisp start to the lesson.

Each unit includes a 'word of the week', 'shape of the week' or 'number of the week' which helps the children to extend their vocabulary and to begin to generate statements about familiar mathematical concepts.

N2 Place-value

Open the box

Ordering 3-digit numbers
Draw a 4 × 2 grid on the board. Cover each space with a blank card, and underneath write a 3-digit number. Choose a child to reveal a number and read it aloud. Check with the class. When all the numbers are revealed the children work in pairs to write all the numbers in order.

Stick there

Place-value in 3-digit numbers
Two dice and cubes
Divide the class into two teams. Write 'H, T and U' on the board for each team. The teams take turns to throw the dice and write the number on the board in the hundreds, the tens or the units column. They are trying to make the largest number they can.
When each team has a 3-digit number they all read it aloud. Which number is larger? That team collects a cube. Continue until one team has seven cubes.

Word of the week

Sample tasks	• How many digits in the number 156? • What is the value of the hundreds digit in 587?
Sample facts	• a number between 10 and 100 has two digits • a million has seven digits • a digit is another word for finger

2

N3 Addition

Grid addition

Addition pairs to 10
Draw a 5 × 2 grid on the board, with a number (1 to 10) in each space. The children copy the grid, writing the number pair to 10. They should try to keep up with you as you write the numbers. When the grid is complete, point to each space in turn, asking a different child to say their answer.

Match me

Addition pairs to 7, 8, 9
Number cards (0 to 9), one set per pair
Select a card at random from a set (0 to 7), hold it up and read it aloud. Children should hold up the number pair to 7.
Repeat for bonds to 8 and 9.

Number of the week 10

Sample tasks	• Say a pair which add to make the number. • Say three numbers which add to make our number. • Say the number which makes our number (10) with 3, 8, 4, ... • Say a pair which differ by our number.
Sample facts	• it is 3 and 7, 4 and 6, ... • it is half of 20, a quarter of 40, ... • it is the first 2-digit number

3

Teacher Cards

The main teaching for the lesson is supported by the Teacher Cards. The front of each Teacher Card provides:

- support for the whole class teaching on the first day of a topic
- a list of the key teaching points addressed by the Unit
- a list of any materials necessary for the teaching
- key vocabulary either introduced or used during the course of the Unit.

N2 Place-value

Teaching points

- To recognise the relationship between hundreds, tens and units
- To partition a 3-digit number into hundreds, tens and units

Materials

- Coins (£1, 10p, 1p)
- Number cards (0 to 9)
- Blu-tack

Key words

- hundreds (H)
- tens (T)
- units (U)
- pence
- penny
- pound
- value

Teaching

- Rehearse the value of each coin. Hold up the 1p coin. *What is its value? One pence, one penny.* Hold up the 10p coin. *How many pennies? Ten. How many 1p coins match it? Ten. Its value is ten pence.* Hold up the £1 coin. *How many pennies? One hundred. How many 10p coins match it? Ten. How many 1p coins match it? One hundred. Its value is one hundred pence or one pound.*

- Write 'H T U' on the board. Remind the children what the letters stand for. Choose a child. Give her three number cards, e.g. 3, 7 and 2. *We are going to make a three-digit number with the cards. Three hundreds.* Stick the 3 under the 'H'. *Seven tens.* Stick the 7 under the 'T'. *Two units.* Stick the 2 under the 'U'. *Three hundred and seventy-two.*

- We shall match the number with coins. Choose three children to be bankers. *You are the hundreds, you are the tens and you are the units. Units are 1p coins, tens are 10p coins, hundreds are £1 coins.*

- Point to the number. *Three hundred and seventy-two. Three hundred.* Encourage the hundreds child to collect three £1 coins. *Seventy, that's seven tens.* Encourage the tens child to collect seven 10p coins. *Two.* The units child collects two 1p coins. *Three hundred and seventy-two.* The children read the number in unison as you point to the cards.

- Collect the cards and coins and repeat several times, using different cards to make different 3-digit numbers. Choose different children to be the bankers.

3 N2

The back of each Teacher Card provides:

- support for further teaching for a subsequent day
- references to differentiated practical activities, including materials and learning outcomes (which are presented in a separate Activity Book)
- guidance on key points, common misconceptions etc., for use during plenary sessions
- references to additional Abacus resources: Mental Warm-up Activities, Activity Book, Textbooks, Photocopy Masters.

N2 Place-value

Materials

- Place-value cards (hundreds, tens and units)

Teaching

- Write a 3-digit number on the board, e.g. 351. Read the number together *Three hundred and fifty-one.*
- Place the hundreds, tens and units cards in three separate piles. Choose three children to take one card each to make the number. Place the cards together to show '351'. Read it together. *Three hundred and fifty-one.* Ask the class *How many hundreds? Three.* Hold up the 300 card. *How many tens? Five tens or fifty.* Hold up the 50 card. *How many units? One.* Hold up the 1 card.
- Ask the three children to hold up the three cards separately. Write '300 + 50 + 1' alongside 351. *Three hundred and fifty and one is three hundred and fifty-one.*
- Repeat for different sets of 3-digit numbers.

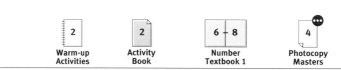

Warm-up Activities · Activity Book · Number Textbook 1 · Photocopy Masters

ACTIVITY 1   *Whole class, in pairs*
Number cards (0 to 9) (PCM 1), place-value boards (PCM 13)
Recognising the value of each digit in a 3-digit number, beginning to compare and order 3-digit numbers

ACTIVITY 2   *2 pairs*
Place-value cards (PCMs 10 to 12)
Partitioning a 3-digit number into hundreds, tens and units, reading and writing 3-digit numbers

ACTIVITY 3   *3 children*
Place-value cards (PCMs 10 to 12), dice (marked 'H', 'T', 'U' on each of two sides), cubes
Recognising the value of each digit in a 3-digit number

ACTIVITY 4   *3-4 children*
Two sets of number cards (0 to 9) (PCM 1), counters
Recognising the value of each digit in a 3-digit number, partitioning a 3-digit number into hundreds, tens and units

ACTIVITY 5   *3-4 children*
Number cards (0 to 9) (PCM 1)
Writing a 3-digit number in words

Plenaries

Key points

- Reciting 3-digit numbers in sequence.
- Recognising and reading 3-digit numbers.
- Recognising that in a 3-digit number, the first digit shows the number of hundreds, the second shows the number of tens, and the third shows the number of units.
- Splitting a 3-digit number into its component parts.

Common difficulty

- Not being able to read 3-digit numbers that contain a zero digit. (Help by making and reading numbers using place-value cards; hundreds and tens only or hundreds and units only.)

Activity

Sit the children in a circle. Choose a 3-digit number, e.g. 185. They take turns to chant the numbers in sequence. Say *Tens* and they count in tens. Say *Hundreds* and they count in hundreds. Say *Ones* and they count in ones. Change the command after a few children by saying *Hundreds, Tens* or *Ones.*

Abacus ❸ Teacher Cards Ginn and Company 1999

The Teacher Cards are divided into two sets: *Number* and *Shape, Data and Measures*. A suggested teaching order, and how this fits with the National Numeracy Framework Planning Grids are given at the end of this book. There are other possible routes through the materials, and the *Shape, Data and Measures* units are presented in a separate block to allow you easily to re-order them.

Starting the lesson

As we know, the way a lesson starts can often dictate the direction of subsequent teaching. If a lesson gets off to a crisp, clear start, the children are likely to be well prepared to engage with the main topic being taught. The National Numeracy Strategy suggests that the first ten minutes of the lesson should be allocated to the practice and reinforcement of those skills and facts which benefit from a 'little and often' approach. In this category are included:

● counting in steps of different sizes, both forwards and backwards
● instant recall of number facts, including discussing ways of remembering the facts that need to be learned by heart
● previously-taught mental strategies.

It is not appropriate to introduce topics for the first time in this mental/oral part of the lesson, and neither should complex operations be explained. It is a time for 'quick-fire' rehearsal of basic skills, for using a strategy that has already been taught, or for choosing an operation to solve a problem quickly.

This part of the session will almost always be number work, although occasionally it may focus on a particular aspect of shape or of measures, such as 'time'. Importantly, it is not necessarily related to the second part of the lesson. For example, a teacher planning a lesson on 2-d shape could (and often would) include a starter activity on number.

This part of the lesson involves working with the whole class and only exceptionally would it be necessary to work with a group or to exempt individuals from the session. The Abacus Mental Warm-up Activities have been specifically written to address this part of the lesson, and provide a wide variety of whole class oral and mental activities for each day's teaching.

Main teaching activity

The function of the main part of the lesson can vary, depending upon various factors (where you are in a topic, the age and maturity of the children etc.). It can provide time for:

● introducing a new topic
● extending previous work and developing children's understanding
● practising a skill or strategy
● using and applying what has been learned
● assessing children's learning
● revising or revisiting a topic.

During the main teaching activity, the teacher can choose how to organise the class. It is quite possible that the lesson you teach on a

Monday will have quite a different style of classroom organisation from the lesson you teach on Tuesday. Clearly, the way you choose to organise the class will be guided by the purpose of the lesson (is it introducing a new topic, or rehearsing an old one?). It will also dictate, to some extent, the teaching strategy you use. For example, a whole class lesson could require a higher percentage of demonstration or explanation, whilst a lesson where groups of children are working together could involve more discussion or description. Taking all these factors into account, on any one day a teacher can choose to organise the class:

- as a whole class session, with direct teaching of the whole class or with a paired investigation
- with the children working in groups, and the teacher working in a focused way with one or two of the groups
- with the children working on a task in pairs or as individuals.

Although the Numeracy Strategy explicitly encourages a flexible approach to the organisation of the main teaching activity, it is clear about three central points:

- the need for teachers to organise the class to fit the purpose of the lesson
- that, although any one lesson may not involve a high proportion of direct teaching of the whole class (or year group) nevertheless, over the complete unit, more than 50% of the teaching is likely to be of this form
- that emphasis should be placed on keeping the children together, so it would not be appropriate to set more than three levels of work for any one year group, and no more than two levels of work per year group in a mixed-age range class (though, clearly, specific provision must be made for particular children who have individual education plans).

It is clear (from the National Numeracy Strategy) that at the heart of the teaching and learning process is the active, direct teaching of a topic. In order to focus on this, teachers need to have:

- a clear objective for each lesson and a set of objectives for the topic
- a definite 'way of teaching' – an image or model (e.g. number line, money) which will be used to represent or demonstrate the mathematics
- a variety of teaching strategies to ensure that the lesson is interactive.

The Abacus Teacher Cards have been designed with these three requirements in mind.

- They provide a clear set of objectives for each unit, with specific learning objectives for each activity.
- They outline the model to be used in teaching the mathematics, and make clear the careful use of vocabulary.
- They also provide a variety of teaching strategies, through the work with the whole class, the group activities, the individual and paired practice and the extra support provided by the Simmering Activities and the Games.

The plenary session

Plenary sessions can often be the hardest part of the lesson to teach successfully. The children (and the teacher!) are tired, and some will require more feedback than others, meaning that boredom may cause disruption. It is hard to plan the plenary session because it will often develop from what has occurred during the lesson.

The purpose of the plenary as explained by the Numeracy Strategy is threefold. It allows time for the following.

- A rehearsal of the main teaching points and summary of the key facts. This is the vital, 'What have we learned today?' part of the lesson.
- Children to present their work to you and others, particularly those who may have been working independently during the main part of the lesson.
- Addressing any difficulties that may have arisen during the lesson.
- Forward planning – helping to make explicit what we are going to do next, and perhaps outlining any homework.

Rehearsing the main teaching points of the lesson is very important for two reasons. Firstly, it allows you to ensure that the learning objectives have been clearly articulated and that children have taken this in. Secondly, it enables you to pick up any difficulties that several children may have shared. Frequently there will be common misconceptions or errors which it can prove useful to address with the whole class. This is not to point out children who have made mistakes, but rather to use common errors as a way of rehearsing the teaching points.

The Abacus Teacher Cards highlight the key learning points and also provide advice on probable common misconceptions which may occur when teaching that particular topic. This will allow you to plan how to address such misconceptions, as well as preparing you to look out for difficulties which are encountered by several children.

Taking feedback from children is difficult to do in a way which does not leave some of the class feeling bored or frustrated. It helps to vary the format of their presentation. For example:

- ask some children to write a question they had to do on the board
- give some children a poster or a strip of paper to record some or all of their work during an activity, so that they can display this
- ask a group to show their work and then choose a question to ask the rest of the class
- choose a pair of children to demonstrate a strategy they have been using
- choose some children to write the answer to a question on the board. The rest of the class have to guess what the question was.

The plenary can also be made easier by discussing a group of children's work with them, and then marking it together just before you bring the whole class together. That group can then show the children their work by showing any pieces you felt were particularly special.

The plenary should end with a suggestion about where the children are going next in mathematics. You may be going to extend the work you did today, or start a new topic. Sometimes you will want to give the class a formal homework task, and on other occasions it will be a more informal activity (e.g. *look out for car numbers that end in '5'. Can you remember one to tell me tomorrow?*).

The plenary should be kept short – a protracted ending will not improve the lesson or make the children learn any more. The lesson needs to be brought to a sharp close, and the plenary enables the teacher to evaluate its success.

Homework

It is important to bear in mind the purposes of homework. A reasonable list might include:

- To involve the parents in their children's learning.
- To help parents keep abreast of what their child can and cannot do.
- To utilise the context of the home and apply some numerical strategies in a non-school situation.
- To encourage parents and children to talk about their number work and for children to explain what they are doing and how.
- To extend the time for learning mathematics and give some extra practice in number work.
- To help parents improve, extend or practise their own numerical strategies.

It is a well-established fact that simply sending home extra mathematics will have little or no effect on many children's attainment, unless effort is made to involve and include their parents in their learning. Therefore, in planning and setting homework, it is important that schools consider the following.

How much and how often?

Little and fairly frequently is better than lots occasionally. It is good to have specific and regular homework 'nights' so that parents and children grow to expect it. An ideal schedule is a shared task at the weekend, so that parents can get involved, and more traditional individual homework once or twice a week, depending on the age of the child.

What type?

Homework is most effective when it is varied – some types of homework might include:

- Shared mathematics tasks, games or other activities where the child discusses the mathematics. The task itself might draw upon the home context.
- Memorisation games or activities to help children with the instant recall of number facts.
- Tasks or activities which help children practise number strategies or skills introduced in school.
- Investigative or problem-solving activities where the child, and the parent, explore an aspect of mathematics together.

The Abacus Homework Books have been planned to include a range of these types of activity. How many parents get involved with the shared activities is not only dependent on the catchment area of the school – research indicates that the attitude of the teachers and the school is a very important factor. By holding meetings, talking to parents, feeding back information and trying to keep them involved, the school can make a huge difference to the number of parents who play an active role in their child's education.

Who is it for?

All children can be encouraged to share mathematics activities with their parents (though the proportion of homework to be done independently will increase as children get older). However, no child should be excluded from an activity because they have difficulty finding a partner to share the task at home. Wherever possible provision for repeating or extending the shared task in the context of school (or at an after school club) should be made.

Assessment

Opportunities for informal assessment occur throughout many of the aspects of an Abacus Unit.

The Mental Warm-up Activities offer opportunities for observational assessment, both for the whole class and for individuals. The continuous feedback in the form of child response to these tasks provides immediate observational measures of performance. The opportunities in this stage of the unit for you to set tasks for varying levels of ability inform further when assessing the feedback. Similarly, during the Plenary sessions, observational assessment opportunities are numerous.

Additional opportunities are provided during the main teaching, both during the interactive element of the teaching, and particularly when the children are engaged in group activities, and you are focusing on a chosen group.

Assessment of the childrens' written work is provided from completed Textbook pages and Photocopy Masters and can be recorded using the Assessment grids on pages 36 to 39.

The Assessment Book offers comprehensive support material for assessing children's progress, within each unit or small set of units. This range includes: a written assessment sheet, oral assessment questions, and suggested sections of the Textbook to use for diagnostic purposes. Also included are common difficulties, and suggested practical activities to help overcome these.

⑥ Differentiation

The National Numeracy Strategy emphasises the need to keep the children together. This links to the requirement that, over a period of time, the proportion of whole class direct teaching should be more than 50% (indeed, it is easy to see how this figure could be closer to 75% in a single year group, given that the mental starter and plenary both involve working with the whole class).

Whole class focus

Although any one lesson may not involve a high proportion of whole class direct teaching, over a complete topic, more than 50% of the teaching is likely to be with the whole class.

Keeping the children together

Emphasis should be placed on keeping the children together, so that it would not be appropriate to set more than three levels of work for any one year group, and no more than two levels of work per year group in a mixed-age class. Clearly, specific provision must be made for particular children who have individual education plans.

This marks a change in attitude to teaching the class, and has repercussions for record-keeping and assessment plans. Given that we are now:
- teaching the same topic to all the children
- practising pre-requisite skills in the mental/oral starter with all the children
- dividing into no more than three groups for the main part of the lesson
it is no longer necessary to write notes on the progress of every single child. The majority of each year group will be engaged in the same tasks, and studying the same mathematics at the same level. For these children, it is important to record deviations from the norm: i.e. what happens if the child was not able to cope with the work given, or if they found it too easy.

It remains as true as ever that children are individuals, who do not all learn or develop at the same rate. The following points may help teachers to cope with the inevitable differences between children when they are teaching the whole class and 'keeping all the children together'.

It is important to remember that the class may be divided into more than three *groups* as long as there are no more than three *levels* of work. Most teachers organise their class so that the children are seated in small groups. They may, therefore, have two or three 'tables' working at the same level, probably on the same activity. One or two more, or less able children may be working on an activity more appropriate for their ability.

Once the topic has been introduced to the children, and the main direct teaching has taken place, it is usual to split the class into groups for most of them to work more independently, and to practise or apply what they have just been taught. It is only in this way that children come to make this knowledge their own. It is also the means of finding out who has, and who has not, grasped the mathematics, and therefore assessing the need for further teaching or practice. It is a sensible strategy to work with the children who need most help first, since this is the group who are likely to need some further teaching before they are able to work independently or semi-independently on the topic.

There is (as the Numeracy Strategy makes clear) a 'hierarchy' of types of work which can be used when planning for groups of children. It is easiest for children to do work which has been laid out for them, and where few decisions have to be made about what is to be done. Thus Textbooks, where the children are given help with laying out their work, and the tasks are still well-defined, are also useful in enabling independent work. Photocopy Masters are also relatively simple for children to use independently from the teacher. Work written on the board is more difficult, since children must lay it out for themselves. Activities involving structural materials, coins, number cards, grids, etc., can require quite a high input from the teacher.

Abacus provides materials for the different levels of work suggested by the Numeracy Strategy and also helps to organise the class so that you are working in a focused way with one group, as recommended. The Activity Book indicates activities targeted at three ability levels, and the Textbooks allow for children who find the work moderately easy to engage with extension activities. It is easiest to organise the class so that those children who are working relatively independently have suitably contained activities – (e.g. Photocopy Masters, Textbooks), whilst those children who have the benefit of your focused attention engage in the activities which enable further directed teaching.

Special needs

Children who have Individual Education Plans (IEPs) will normally be included in the whole class parts of the lesson, such as the mental/oral starter, the plenary, and the introduction to the topic in the main teaching activity. It will be rare for a child to be excluded from these parts of the lesson, although behaviour management, or their skills level, may mean that they are only present for some of these activities. Once a topic has been introduced, the teacher will need to ensure that the follow-up activities are at a level which is appropriate for the child and which fit with their development plan. Normally children with IEPs will work on the same topic as the remainder of the class, but at a lower level or by addressing some of the pre-requisite skills needed to engage with that topic. This work can be selected either from the Abacus Unit itself, drawing upon appropriate activities, or from appropriate sections of the Numeracy Support Book.

Multiplication and division

In Year 3 the concepts of multiplication and division are consolidated through the arrangement of objects in arrays, from which the commutative property (i.e. $3 \times 4 = 4 \times 3$) is recognised and emphasised. The $\times 3$ and $\times 4$ tables are introduced and mental recall of the associated multiplication and division facts is encouraged.

The 'grouping' concept of division is developed using the groups of objects to illustrate the concept of multiplication. In this way, division and multiplication are developed together.

The skill of recalling double facts is further developed, and halving is introduced as the reverse of doubling.

Later in Year 3 pupils are introduced to the multiplication of 1-digit numbers by both 10 and 100, as well as the multiplication of 2-digit numbers by 10.

Money

Abacus uses money to provide a 'real life' and familiar context within which to locate many numerical operations and concepts. Children are encouraged to see 'hundreds, tens and units' in terms of pound coins, ten pence pieces and one pence pieces. All children will regard money and coins as both interesting and important. Its use as a structured model for place-value capitalises on this interest.

Children see adults using, enjoying and worrying about money. They are well aware of the importance accorded to money in our society. They know that money transactions are an essential part of our daily lives. Unlike coloured bricks, blocks or interlocking cubes, money has an importance and a reality outside the primary classroom. It is this fact that makes money such a powerful teaching tool. Children want to operate using money. They want to understand how money works, and to participate in the adult transactions that they see around them.

Measures

In all aspects of measures (e.g. length, weight, capacity, angle), Abacus provides pupils with activities which develop the important skill of estimation. Providing children with estimating experiences enhances their understanding of the measurement concepts involved.

Weight and mass

Throughout this stage of work in Abacus the work 'weight' is used in preference to 'mass'. Scientifically, the weight is the force exerted by the earth's gravitational field on an object: weight can then be crudely interpreted as the pull towards the ground. The mass of an object is the quantity of matter it contains. The force of attraction varies with the position of the object, but the quantity of matter remains fixed. So, for example, an astronaut weighs less on the Moon because the Moon's gravitational attraction is weaker than the Earth's. The astronaut's mass – the amount of matter in his body – is unchanged.

The weight of an object can be found by using a spring balance, bathroom scale or direct-reading kitchen scale, but the mass has to be found by using balance scales where an object is placed in one pan and a balance obtained by placing objects of known mass in the other.

Though this distinction between weight and mass is clear to the scientist, it is rarely made in daily life. Later in Abacus, when units have been developed and understood, this distinction will be relatively easy to grasp, but at this stage the word weight is used in accordance with its everyday usage. If you are concerned about this decision, the word mass may be used when it is obviously more appropriate, but the intention is that no undue worry should be centred around the distinction between the two words.

2-d and 3-d Shapes

The word 'vertex' is introduced to replace the word 'corner' introduced at Key Stage 1. The correct terminology is 'sides' and 'vertices' for 2-d shapes, and 'faces', 'edges' and 'vertices' for 3-d shapes.

The study of 2-d shapes is developed to include the term 'quadrilaterals'

Regular and irregular 2-d shapes

A pentagon is a five-sided polygon (a 2-d shape with straight sides). Pentagons can be regular (all sides are the same length, and all angles the same size), or irregular (all sides are not the same length, nor angles the same size). Pentagons found in packs of 2-d shapes tend to be regular.

It is not necessary, at this stage, for pupils to distinguish between the two types, but it is important for them to see both regular and irregular pentagons, and recognise all five-sided polygons as pentagons. The same is true of hexagons, octagons etc.

Consequently, the Teacher Cards and Textbook pages include both regular and irregular polygons.

Prisms

In the study of 3-d shapes, pupils are introduced to the concept of a prism. A prism is a 3-d shape whose cross-section is always the same shape. Consequently, the two end-pieces are this same shape. If the shape of the end-piece is a triangle, the prism is called a triangular prism. If is a pentagon, then it is called a pentagonal prism etc.

If the end-piece is a square or rectangle, then the prism might be a cube or a cuboid. So cubes and cuboids are special types of prism.

The concept of 'cross-section' is a difficult one for pupils at this stage. Since many packs of cheese are prism-shapes, the idea of the same shaped cross-section can be illustrated by slicing though packs of cheese.

Angle

In Year 3, the pupils' concept of 'direction' is introduced using the four-point directional compass, and this is used to enhance the concept of angle as a measure of turn, using right angles.

Position

Position on a grid can be located in two different ways, and it is important to distinguish between them. If the horizontal and vertical lines of the grid are labelled, then positions are located by the point where two lines meet. If the spaces between the grid lines are labelled, then positions are located by the grid cell where the two spaces meet.

In Abacus 3 we focus on locating position by labelling the spaces. In Abacus 4, the idea of locating position is extended to identifying points (co-ordinates).

Graphs

Abacus requires that all graphs are titled and have labelled axes. Usually, the horizontal axis represents the item (e.g. types of sport), and the vertical axis represents the frequency (e.g. number of votes).

In a block graph (introduced in Abacus 1 and 2), the numbering of the frequencies on the vertical axis should be written in the spaces between the blocks.

A bar graph (introduced in Abacus 3) is similar to a block graph, but uses a bar to replace each tower of blocks. In a bar graph, the numbering on the vertical axis should be written directly alongside the divisions. Also, items on the horizontal axis, should be separated, so that there are spaces between the bars.

In the Activity Book, there is a blank bar graph, on which pupils can draw their graphs accurately.

Framework for teaching mathematics matching Chart

3

Numbers and the number system

Counting, properties of numbers and number sequences

Count larger collections by grouping them: for example, in tens, then other numbers.	N7
Describe and extend number sequences: count on or back in tens or hundreds, starting from any two- or three-digit number;	N7, N8, N21, N30, N36, N42
count on or back in twos starting from any two-digit number, and recognise odd and even numbers to at least 100;	N22, N36
count on in steps of 3, 4 or 5 from any small number to at least 50, then back again.	Warm-ups
Recognise two-digit and three-digit multiples of 2, 5 or 10, and three-digit multiples of 50 and 100.	Warm-ups, N21

Place-value and ordering

Read and write whole numbers to at least 1000 in figures and words.	N1, N21
Know what each digit represents, and partition three-digit numbers into a multiple of 100, a multiple of ten and ones (HTU).	N1, N2, N15, N18
Read and begin to write the vocabulary of comparing and ordering numbers, including ordinal numbers to at least 100.	Throughout
Compare two given three-digit numbers, say which is more or less, and give a number which lies between them.	N2, N15
Say the number that is 1, 10 or 100 more or less than any given two- or three-digit number.	Warm-ups, N1, N8, N15
Order whole numbers to at least 1000, and position them on a number line.	N1, N15, N29

Estimating and rounding

Read and begin to write the vocabulary of estimation and approximation.	N7, N12, N28, M1, M5
Give a sensible estimate of up to about 100 objects.	N7, Warm-ups
Round any two-digit number to the nearest 10 and any three-digit number to the nearest 100.	N29

Fractions

Recognise unit fractions such as $\frac{1}{2}, \frac{1}{3}, \frac{1}{4}, \frac{1}{5}, \frac{1}{10}$ and use them to find fractions of shapes and numbers.	N12
Begin to recognise simple fractions that are several parts of a whole, such as $\frac{3}{4}, \frac{2}{3}$ or $\frac{3}{10}$.	N28, N41
Begin to recognise simple equivalent fractions: for example, five tenths and one half, five fifths and one whole.	N41
Compare familiar fractions: for example, know that on the number line one half lies between one quarter and three quarters.	N41
Estimate a simple fraction.	N12, N28

Calculations

Understanding addition and subtraction

Extend understanding of the operations of addition and subtractions, read and begin to write the related vocabulary, and continue to recognise that addition can be done in any order. Use the +, and – and = signs.	N4, N33, throughout
Extend understanding that more than two numbers can be added, add three or four single-digit numbers mentally, or three or four two-digit numbers with the help of apparatus or pencil and paper.	N4, N34
Extend understanding that subtraction is the inverse of addition.	N17

30

Rapid recall of addition and subtractions facts

Know by heart:	
all addition and subtraction facts for each number to 20;	N3, N4, N16
all pairs of multiples of 100 with a total of 1000 (e.g. 300 + 700).	Warm-ups
Derive quickly: all pairs of multiples of 5 with a total of 100 (e.g. 35 + 65).	N13, N14, N23, N30, N33

Mental calculation strategies (+ and –)

Use knowledge that addition can be done in any order to do mental calculations more efficiently. For example:	
put the larger number first and count on;	N30, N33
add three or four small numbers by putting the largest number first and/or by finding pairs totalling 9, 10 or 11;	N4, N30, N33
partition into '5 and a bit' when adding 6, 7, 8 or 9 (e.g. 47 + 8 = 45 + 2 + 5 + 3 = 50 + 5 = 55);	N16, N19
partition into tens and units, then recombine (e.g. 34 + 53 = 30 + 50 + 4 + 3).	N18, N19, N20
Find a small difference by counting up from the smaller to the larger number (e.g. 102 – 97).	N17, N32
Identify near doubles, using doubles already known (e.g. 80 + 81).	Warm-ups, N19, N20
Add and subtract mentally a 'near multiple of 10' to or from a two-digit number by adding or subtracting 10, 20, 30 and adjusting.	N19, N20, N24
Use patterns of similar calculations.	Warm-ups
Say or write a subtraction statement corresponding to a given addition statement, and vice versa.	N17
Use known number facts and place value to add/subtract mentally.	N5, N6, N23, N31, N34
Bridge through a multiple of 10, then adjust.	N6, N16, N17

Pencil and paper procedures (+ and –)

Use informal pencil and paper methods to support, record or explain HTU ± TU, HTU ± HTU.	N30, N34, N35, N42
Begin to use column addition and subtraction for HTU ± TU where the calculation cannot easily be done mentally.	N42, N43

Understanding multiplication and division

Understand multiplication as repeated addition.	N9
Read and begin to write the related vocabulary.	Throughout
Extend understanding that multiplication can be done in any order.	N26, N38
Understand division as grouping (repeated subtraction) or sharing. Read and begin to write the related vocabulary.	N9, N40
Recognise that division is the inverse of multiplication, and that halving is the inverse of doubling.	N9, N10, N11, N26, N39, N40
Begin to find remainders after simple division.	N40
Round up or down after division, depending on the context.	N40

Rapid recall of multiplication and division facts

Know by heart:	
multiplication facts for the 2, 5 and 10 times-tables.	N10, N25
Begin to know the 3 and 4 times-tables.	N27, N39
Derive quickly:	
division facts corresponding to the 2, 5 and 10 times-tables;	N10, N25, Warm-ups
doubles of all whole numbers to at least 20 (e.g. 17 + 17 or 17 × 2);	N11, N24, Warm-ups
doubles of multiples of 5 to 100 (e.g. 75 × 2, 90 × 2);	N24, Warm-ups
doubles of multiples of 50 to 500 (e.g. 450 × 2);	Warm-ups
and all the corresponding halves (e.g. 36 ÷ 2, half of 130, 900 ÷ 2).	N11, N24, Warm-ups

Mental calculation strategies (× and ÷)

To multiply by 10/100, shift the digits one/two places to the left.	N37
Use doubling or halving, starting from known facts (e.g. 8 × 4 is double 4 × 4).	N11, N24, N39
Say or write a division statement corresponding to a given multiplication statement.	N27, N31, N39
Use known number facts and place value to carry out mentally simple multiplications and divisions.	N38, N39

Checking results of calculations

Check subtraction with addition, halving with doubling and division with multiplication.	Throughout
Repeat addition or multiplication in a different order.	Throughout
Check with an equivalent calculation.	Throughout

Solving problems

Making decisions

Choose and use appropriate operations (including multiplication and division) to solve word problems and appropriate ways of calculating: mental, mental with jottings, pencil and paper.	Throughout

Reasoning about numbers or shapes

Solve mathematical problems or puzzles, recognise simple patterns and relationships, generalise and predict. Suggest extensions by asking 'What if?'	Throughout
Investigate a general statement about familiar numbers or shapes by finding examples that satisfy it.	Throughout
Explain methods and reasoning orally and, where appropriate, in writing.	Throughout

Problems involving 'real life', money and measures

Solve word problems involving numbers in 'real life', money and measures, using one or more steps, including finding totals and giving change, and working out which coins to pay.	Throughout
Explain how the problem was solved.	Throughout
Recognise all coins and notes. Understand and use £.p notation (for example, know that £3.06 is £3 and 6p).	Throughout

Handling data

Organising and using data

Solve a given problem by organising and interpreting numerical data in simple lists, tables and graphs, for example: Simple frequency tables; pictograms – symbol representing two units; Bar charts – intervals labelled in ones then twos; Venn and Carroll diagrams (one criterion).	D1, D2, D3, D4, D5

Measures, shape and space

Measures

Read and begin to write the vocabulary related to length, mass and capacity.`	M1, M2, M5, M7
Measure and compare using standard units (km, m, cm, kg, f, l, ml), including using a ruler to draw and measure lines to the nearest half centimetre.	M1, M2, M5, M7
Know the relationships between kilometres and metres, metres and centimetres, kilograms and grams, litres and millilitres.	M1, M2, M5, M7
Begin to use decimal notation for metres and centimetres.	M1
Suggest suitable units and measuring equipment to estimate or measure length, mass or capacity.	M1, M7
Read scales to the nearest division (labelled or unlabelled).	Throughout, M1, M2, M5, M7
Record estimates and measurements to the nearest whole or half unit (e.g. 'about 3.5kg'), or in mixed units (e.g. '3m and 20cm').	M1, M5
Read and begin to write the vocabulary related to time.	M3, M4, M6, M8
Use units of time and know the relationships between them (second, minute, hour, day, week, month, year). Suggest suitable units to estimate or measure time.	M3, M6, M8
Use a calendar. Read the time to 5 minutes on an analogue clock and a 12-hour digital clock and use the notation 9:40.	M3, M4, M8

Shape and space

Classify and describe 3-D and 2-D shapes, including the hemisphere, prism, semi-circle, quadrilateral referring to properties such as reflective symmetry (2-D), the number or shapes of faces, the number of sides/edges and vertices, whether sides/edges are the same length, whether or not angles are right angles	S1, S4, S5, S6
Make and describe shapes and patterns: for example, explore the different shapes that can be made from four cubes.	S1, S4, S6
Relate solid shapes to pictures of them.	S1, S5, S6
Identify and sketch lines of symmetry in simple shapes, and recognise shapes with no lines of symmetry.	S2
Sketch the reflection of a simple shape in a mirror line along one edge.	S2
Read and begin to write the vocabulary related to position, direction and movement: for example, describe and find the position of a square on a grid of squares with a rows and columns labelled.	S8
Recognise and use the four compass directions N, S, E, W.	S3, S7
Make and describe right-angled turns, including turns between the four compass points.	S3, S6, S7
Identify right angles in 2-D shapes and the environment.	S6
Recognise that a straight line is equivalent to two right angles.	S6, S7
Compare angles with a right angle.	S6, S7

⑩ Classroom materials

Materials provided in the *Resource Bank*

Number line (0 to 20) – large, double-sided, full-colour number cards, for demonstration

Number line (0 to 100) – medium, double-sided two-colour number cards, for demonstration

Class pack of small number cards (0 to 10) – 30 sets, for individual children

Group pack of small number cards (0 to 10) – 10 sets for group work

Group pack of small number cards (0 to 30) – 5 sets for group work

Pack of small number cards (0 to 100) – 1 set for group work

Wall chart pack (0 to 99 grid and 1 to 100 grid) – 6 sets for a school

Place-value cards (standard pack) – 5 sets of units, tens and hundreds

Place-value cards (extended pack) – 5 sets of units, tens, hundreds, thousands, tenths and hundredths.

Other specialised resources, such as small number grids and number tracks, are included as Photocopy Masters in the Activity Book.

Games pack

8 games referenced to specific Teacher Card Units.

Assumed mathematical materials in the classroom

Interlocking cubes

Counters

Spotty dice (1 to 6)

Numbered dice (1 to 6) and (0 to 5) and (1 to 10)

Calculators

Dominoes

Playing cards

Coins (real and plastic)

Base Ten equipment

Sets of plastic 2-d shapes

Sets of solid 3-d shapes

Plastic mirrors

Containers (various, for Capacity)

Analogue clock with moveable hands

Digital clock

Sand timers or rocking timers

Seconds timer or stopclock

Weighing balances/weighing scales/ kitchen scales

Set of weights

Rulers/metre sticks/tape measures

Centicubes

Pegboards and pegs

Direction compass

Other materials

Washing line with pegs attached

Hoops

Cloth (Feely) bags

Blu-tack

Pasta shapes/dried beans/lentils

Soft toys

3-d model-making equipment e.g. Polydron, Clixi, ...

Coat-hanger and keyrings

Post-it notes

Plastic spider

Exemplar planning grid: autumn

Note: In the grid, the Abacus Unit refers to the main teaching focus of the lesson. Mental and oral skills, such as counting, doubling, addition pairs etc. will be practised every day, using the Mental Warm-up Activities.

Unit	Topic	Abacus Unit	Teaching points	Notes
1	Place-value, ordering estimating, rounding Reading numbers from scales	N1 Place-value	To read and write numbers to at least 1000 To recognise the value of each digit in numbers up to 1000	
		N2 Place-value	To recognise the relationship between hundreds, tens and units To partition a 3-digit number into hundreds, tens and units	
2–3	Understanding + and – Mental calculation strategies (+ and –) Money and 'real life' problems	N3 Addition	To rehearse addition facts to 10 and learn those to 20 To recognise related addition facts	
		N4 Addition	To extend understanding of the concept of addition, relating to money To recognise that addition can be done in any order (4 + 8 + 16) To rehearse addition facts to 20 (20p)	
	Making decisions and checking results	N5 Addition	To use known facts to add To recognise when units add to 10 when adding a 1-digit number to a 2-digit number (e.g. 24 + 6)	
		N6 Subtraction	To subtract a 1-digit number from 2-digit numbers using known facts (21 – 5) To recognise when a subtraction leaves a multiple of ten (e.g. 24 – 4)	
4–6	Measures, including problems	M1 Length	To estimate and measure lengths in centimetres and metres To introduce the kilometre and recognise its relationship with metres	
	Shape and space Reasoning about shapes	M2 Length	To recognise the relationship between metres and centimetres To find the total of two lengths To find the difference between two lengths	
		S1 2-d shape	To rehearse the names of common 2-d shapes (square, rectangle, triangle, pentagon, hexagon, octagon, circle) To recognise quadrilaterals as shapes with four straight sides	
		S2 Symmetry	To recognise line symmetry in shapes To recognise shapes with no lines of symmetry	
		S3 Direction	To recognise the four-point compass directions (N, S, E, W)	
7	Assess and review			

43

Unit	Topic	Abacus Unit	Teaching points	Notes
8	Counting and properties of numbers Reasoning about numbers	N7 Numbers to 1000 N8 Numbers to 1000	To count in ones and tens, including from and to numbers above 100 To count objects, grouping in tens or fives To rehearse counting in ones and tens, starting at any 3-digit number To count in hundreds, starting at any 3-digit number To say the number ten or one hundred more or less than any 3-digit number	
9–10	Understanding × and ÷ Mental calculation strategies (× and ÷) Money and 'real life' problems Making decisions and checking results	N9 Multiplication/ division N10 Multiplication/ division N11 Multiplication/ division	To rehearse the concept of addition To recognise multiplication as repeated addition To recognise division as repeated subtraction To begin to recognise the relationship between multiplication and division. To rehearse the ×2 multiplication table To recall the ×2 multiplication facts To rehearse division by 2 as grouping in twos To recognise division as the reverse of multiplication To rehearse doubling as twice a number To recognise halving as the reverse of doubling	
11	Fractions	N12 Fractions	To rehearse recognition of unit fractions, e.g. $\frac{1}{2}$, $\frac{1}{3}$, $\frac{1}{4}$ To find one half, one quarter, one third of amounts	
12	Understanding + and – Mental calculation strategies (+ and –) Time, including problems Making decisions and checking results	N13 Addition/ subtraction N14 Addition M3 Time M4 Time	To add and subtract multiples of ten To add and subtract multiples of five To use known facts to solve problems in length and money To rehearse pairs of multiples of ten which make 100 To begin to learn pairs of multiples of 5 which make 100 (using length and money) To tell the time to the quarter hour on analogue and digital clocks To recognise five minute intervals on analogue and digital clocks To calculate time intervals in units of five minutes. To tell the time to the nearest five minutes on analogue and digital clocks To recognise a time a number of minutes earlier or later than a given time To solve simple problems involving five minute intervals	
13	Handling data	D1 Tally charts	To construct and interpret a tally chart To collect data	
14	Assess and review			

Exemplar planning grid: spring

Note: In the grid, the Abacus Unit refers to the main teaching focus of the lesson. Mental and oral skills, such as counting, doubling, addition pairs etc. will be practised every day, using the Mental Warm-up Activities.

Unit	Topic	Abacus Unit	Teaching points	Notes
1	Place-value, ordering, estimating, rounding	N15 Place-value	To compare two 3-digit numbers, recognising which is more and which is less To say numbers which lie between two 3-digit numbers To say a number which is 1, 10, 100 more/less than a 3-digit number	
2–3	Understanding + and – Mental calculation strategies (+ and –)	N16 Addition	To rehearse addition pairs to 10/20 To add two numbers bridging across 10 $(35 + 7 = 35 + 5 + 2)$ To solve money problems $(34p + ? = 40p, 34p + ? = 50p)$	
	Money and 'real life' problems Making decisions and checking results	N17 Addition/subtraction	To rehearse addition pairs to 10 To find small differences between 2-digit numbers bridging across 10 $(33 - 27 =)$	
		N18 Money	To recognise and use £ and p notation To add and subtract amounts of money (multiples of ten from £1.42)	
		N19 Addition/subtraction	To add/subtract multiples of 10 to/from 2/3-digit numbers To add/subtract 2-digit numbers (looking for multiples of 10) $(64 - 42 = 64 - 40 - 2)$	
		N20 Addition/subtraction	To add/subtract multiples of 10 to/from 2/3-digit numbers To add/subtract near multiples of 10 $(46 - 29)$	
4–6	Shape and space Reasoning about shapes	S4 3-d shape	To recognise and name prisms To understand the properties of a prism	
		S5 3-d shape	To rehearse the names of 3-d shapes (cube, cuboid, cylinder, cone, pyramid, prism, sphere) To sort 3-d shapes based on numbers of faces, edges and vertices	
	Measures, and time, including problems	M5 Capacity	To measure capacity using millilitres To recognise the relationship between litres and millilitres To read and interpret scales	
		M6 Time	To recognise the relationships between units of time: seconds, minutes, hours, days	
7	Assess and review			

45

Wakefield

Joining in

Becoming a part of Champion Coaching

During April and May each year the Sports Councils invite local authorities to national and regional briefings with staff from Champion Coaching. The purpose of the briefings is to help these potential partners understand the necessary commitment to Champion Coaching before they apply. As the application document states:

Champion Coaching is not for the weak of heart. The National Coaching Foundation believes strongly in helping children access quality coaching opportunities, in raising the profile of coaching and in contributing to the health of school-age sport. There is an enormous amount of work to be accomplished in this area.

After the briefings, local authorities are encouraged to apply. During the application process they are asked to examine their philosophy and existing provision for youth sport. They do this by comparing it with a schedule of 'readiness factors'.

Community readiness factors

We have identified these readiness factors to help communities assess their existing structures for youth sport. They are the factors that contribute significantly to the productive design, development and delivery of quality-assured sport programmes for children.

Champion Coaching operates most effectively when integrated into a comprehensive development programme. Comparing the current provision with the readiness factors indicates how much of the total development programme is already in existence before the introduction of Champion Coaching. Inevitably this will vary enormously from scheme to scheme, and it is unlikely that any one scheme will have all of the factors in place.

Identifying omissions is an important process, not only for the schemes that hope to join, but also for the schemes that are already working with Champion Coaching and trying to improve their overall youth sport and coaching development policy. These schemes have already made a commitment to put the omissions right, and new schemes take on this responsibility too.

1 The planning framework

Many different agencies and organisations have a legitimate involvement with the provision of youth sport, but the one overriding problem is lack of co-ordination. This is best remedied by a planning framework which may take a variety of forms, including:

● Local Coach Development Strategies (see point 8 below)

● Governing bodies' Youth Sport Development Plans

● Focus Sport Development Plans

● Youth Sport Strategies as part of Regional Recreation Strategies

● Local authority District Sport and Recreation Plans

2 The presence of a Youth Sport Manager

Central to every Champion Coaching scheme is the Youth Sport Manager, who will:

● liaise with the National Coaching Foundation in the local design and development process

● co-ordinate the community agencies involved in providing youth sport, so that the programmes are delivered efficiently

● organise and oversee the use of the Champion Coaching Kit (see Chapter 4) for community members including coaches, teachers, families, club personnel and school officials

● explain and demonstrate the Champion Coaching philosophy to the network of community agencies.

No other single factor contributes to the success of a Champion Coaching scheme as much as the commitment of the Youth Sport Manager. The right person, with sufficient time and support from senior management, is essential. So such a person must be already in place (quite likely with a different job title), or at least identified and available.

3 Effective networking

The key to successful implementation of development plans is co-ordination on the ground. For a scheme to be successful there must be some evidence of existing networking between agencies. The Sandwell Sport Development Unit based within the local education department is a good example (see pages 134–135). Effective networking is demonstrated by partnership in the delivery of current sport programmes, and by the existence of sport development units. Where staff are dedicated to intervention and outreach work with networking as a central philosophy, co-ordination on the ground is dynamic and productive.

4 Existing school–community links

If there is a rapport and mutual respect between education and leisure services, the planning and operation of Champion Coaching sport programmes is smoother, less costly and more fully embraced by both.

5 Partnership posts

Partnership posts link relevant agencies together. They include school–community liaison officers, co-funded sport development officers and sport-specific development officers for youth sport. If a community possesses one or more of these officers they can have a positive role within Champion Coaching and are in a good position to help develop a co-ordinated structure. These posts already exist in over 20 schemes.

Facing page: Netball at the Sandwell scheme — managed from within the local education department

6 Community assets

Champion Coaching sport programmes require a number of community assets. The list of ideal requirements includes:

● Motivated children
● Senior management support for youth sport
● Qualified and appropriate coaches
● Co-operative teachers and schools
● Safe child-friendly facilities
● Existing clubs with potential for junior sections
● Interest from various other agencies (including the volunteer sector)

It is not essential for all of these to be in place beforehand, but any that are lacking should be included in future planning.

7 Previous history of long-term schemes

It is important to distinguish communities who are looking only for a short-term benefit, from those who genuinely want a blueprint for their own sport development and who may need financial assistance to start the process. One indicator may be their previous history of entering into a long-term scheme. Communities who see Champion Coaching merely as an incentive to deliver a year's worth of sport programmes are doomed to failure.

However, when they view Champion Coaching as a high-profile means of establishing their own long-term sport development, there is a greater chance of success and sustainability.

8 Implementation of a Local Coach Development Strategy

So much of Champion Coaching depends on identifying, recruiting, deploying and evaluating appropriate coaches for children, that a local coaching strategy is highly desirable. This is a two-way process: an existing Local Coach Development Strategy can help prepare the way for Champion Coaching, while the process of delivering Champion Coaching sport programmes can clarify the stages needed for devising and implementing such a strategy. National Coaching Foundation Coaching Development Officers can help in this process.

9 A compatible philosophy

Champion Coaching has a philosophy about the value of sport for young people, including the responsibility of everyone involved to focus on fair play. This basic philosophy must be shared by the potential partner if mutual goals are to be set and achieved. This is therefore one of the few absolute prerequisites for joining Champion Coaching.

10 Commitment — the extra factor

The commitment to build a better sporting future for the community's children through quality-assured coaching is of decisive importance. This commitment, plus the determination to put it into practice, can be one of the largest components of success.

****

The factors listed above are not exhaustive, but are intended to convey the sort of aims to which partner agencies should be committed. It is important for them to recognise that these factors are vital elements in the building of any long-term structures for youth sport, whether linked to Champion Coaching or not. If Champion Coaching can act as a catalyst to spur their development, significant progress will have been achieved.

Pointing the way forward

All applications are successful in one way or another. Even if a scheme is turned down because the community is not yet ready, the process of applying can have valuable results in itself. As Jeremy Harries, Sports Council Liaison Officer for the North West Region, states:

For some local authorities in this region Champion Coaching has provided a catalyst for action. A desire to be involved has provoked a lot of thought and activity on their part, although it has not necessarily resulted in their pursuit of further involvement. For some it has provided a framework for their own action.

And for those local authorities that are accepted to become part of Champion Coaching it is the start of an exciting period of development: they have the satisfaction of joining a growing movement in youth sport and coaching as they implement the Champion Coaching Blueprint.

Chapter 4

Windsor and Maidenhead

Vital resources

Supporting YSMs and coaches in the community

Youth Sport Managers

- Champion Coaching Guide
- Introductory OHPs

Coach education

- Coach Briefing
- Coach Workshop
- Senior Coach Briefing
- Senior Coach Workshop

THE CHAMPION COACHING KIT

REACHING OUT TO THE COMMUNITIES

CHAMPION COACHING
CHAMPION COACHING
CHAMPION COACHING
CHAMPION COACHING

- Young People in Sport Workshop
- Headteachers and Governors Workshop
- Junior Club Workshop
  + Taking ideas back to your club
- Teacher Liaison
- Newsletters

To inform the wider community

- Head Coach Guide
- Coach Log
- Sport Log
- Using the Sport Log

To support sport programmes

The expansion of Champion Coaching has led to a distinct change in the way communities were made aware of what was happening and also in the way coaches were trained. In 1991 a few senior staff had travelled the country meeting parents, coaches and teachers to help them understand the message of Champion Coaching. It would have been impossible to do the same with schemes, and it would also have been undesirable.

The mission statement declares that we work 'within a co-ordinated community structure', and it was soon obvious that the best way to deliver the Champion Coaching message is: 'in the community and by the community'. We needed material to help everyone to do this, so a comprehensive package was devised.

Although the Blueprint forms a very substantial basis, there are three other major elements in the Champion Coaching package:

● The Champion Coaching Kit
● The National Coaching Foundation Scholarship
● Youth Sport Manager seminars.

The Champion Coaching Kit

The Champion Coaching Kit is a set of resources that can be used by the Youth Sport Managers to influence the communities in which they work. In sport the word 'kit' conjures up visions of tracksuits and team shirts, the things you put on before you go and play. The Champion Coaching Kit has a similar function. It puts those who use and receive it in the right frame of mind to help young people in sport. It is part of their preparation: when they try it on, and it feels right, they perform much better.

The Kit has four parts — resources for:

● Youth Sport Managers
● Informing the wider community
● Coach education
● Supporting sport programmes.

Material to help with presentation is an important part of the Champion Coaching Kit.

Resources for Youth Sport Managers

In order to deliver the Blueprint, Youth Sport Managers need a comprehensive guide to help them. The *Champion Coaching Guide* was developed to do just that, and contains the following sections:

● The Blueprint
● The major partners in Champion Coaching
● Networking
● Choosing sports
● Exit routes
● Working with coaches
● Choosing the children
● Monitoring and evaluation
● Communication.

The *Guide* has been praised by Youth Sport Managers as a simple and effective explanation of the concept of Champion Coaching and of how to put it into practice. As Fiona Fortune, YSM from Coventry, said:

Where would I be without my Guide? It has been really useful and answered nearly all of my questions on Champion Coaching.

To make the *Guide* even more effective, a 'yellow pages' section is included. This contains addresses of:

● Youth Sport Managers
● National Coaching Foundation Coaching Development Officers
● National governing body liaison officers
● Sports Council liaison officers.

It also contains detailed information about the proposed S/NVQ-equivalent qualifications for coaching in all sports.

Youth Sport Managers often have to inform their own organisations about Champion Coaching, so we have provided a set of introductory OHPs and speaking notes to help them explain the concept of Champion Coaching in a straightforward way.

Informing the wider community

Young people in sport are strongly influenced by five groups:

● Parents
● Coaches
● Teachers
● Junior club personnel
● Headteachers and school governors.

The challenge was twofold: to provide material for community use with these groups, and to train the people who were going to use it.

Teachers

Champion Coaching believes in using teachers' knowledge to help choose appropriate sports, shape the sport programmes, and nominate children for them. Youth Sport Managers liaise with teachers in the early stages of planning, to make this process as effective as possible. Two 'teacher briefings' have been developed to assist the YSMs. They cover the important areas of:

- potential leadership roles for teachers within Champion Coaching
- criteria for nominations
- fair play and sporting attitudes
- scholarships for coaches (see page 28)
- the selection process.

Parents

When parents have children involved in sport they soon become aware of their responsibilities. The Young People in Sport Workshops, delivered by Youth Sport Managers, raise some of the issues for parents of children in sport, and give them the chance to discuss them with other parents. The workshop syllabus covers:

- Why children take part in sport
- What we think children gain from sport
- How parents and coaches can help children
- How to develop fair play.

These workshops are also opportunities for YSMs to inform parents about all the sporting opportunities in their local community.

Junior clubs

An essential part of Champion Coaching is providing an exit route for each child. This exit route is usually to a local club with an established junior section. The Junior Club pack helps clubs understand the needs of young people in sport. It covers the following topics:

- Why clubs need young people
- How clubs can prepare to accept youngsters
- Making an action plan for a junior section.

This important pack is delivered at workshops for local clubs by National Coaching Foundation Development Officers because of their knowledge of local coaching and their ability to help clubs provide coach education. Those taking part are given the chance to discuss their ideas with each other, and at the end of the workshop they are sent away with an information sheet entitled *Taking ideas back to your club*.

Headteachers and governors

As much of the groundwork for Champion Coaching takes place in schools, it is important for the management to be fully informed, especially concerning Champion Coaching's potential for encouraging community use of school facilities. The Headteachers and Governors Workshops give them the opportunity to analyse the existing situation of sport in

Parents and players in Clwyd

the extended curriculum, and offer them some models for community use. The sections include:

- The partnership approach
- The advantages of linking with the community
- Action for headteachers and governors.

The workshops are delivered by a group of specially trained tutors who include:

- Directors of leisure services
- Advisers for PE
- Heads of sport development
- Other headteachers
- Education officers.

Newsletters

So much happens in Champion Coaching that we always have lots of news. The newsletter comes out six times a year and serves to inform the wider community of our current work in Champion Coaching. It gives us a chance to tell the story and to be proud of our achievements.

Coach education

During the Champion Coaching pilot project a substantial coaching package was offered to head coaches; this included a weekend course in Loughborough with some of the best people in UK coach education. Expansion in 1993 meant shifting the coach

education into the communities, with some pleasing results:

- All coaches at proposed equivalent S/NVQ levels 2, 3 and 4 have been offered a coach education package.
- NCF Coaching Development Officers were able to expand their coach education into scheme areas.
- Schemes have combined to offer NCF and national governing body courses.
- Head coaches have supported coaches through 'mentoring'.
- Many coaches took NCF courses for the first time.
- Many coaches took NCF courses in addition to those required.

Coach education and support provided at each level is as follows:

Coach (proposed equivalent S/NVQ level 2)

Coaches are welcomed to Champion Coaching with a briefing from the Youth Sport Manager. They then have a workshop with the head coaches to plan the sport programme in detail. This includes:

- How to develop fair play, and support good sporting attitudes in children
- Reviewing safety procedures
- Detailed planning of each session
- Some goal-setting for the coaches.

Senior Coach (proposed equivalent S/NVQ level 3)

In most cases these are the head coaches in the sport programmes. They receive a briefing from the YSM outlining their role in Champion Coaching, and are given the *Head Coach Guide* and *Coach Log* as support material.

Each senior coach attends a 4-hour workshop which helps them prepare for their possible role as a mentor to support and guide those coaches working with them. The workshops have been highly successful and significant. Most governing bodies realise that senior coaches acting as mentors are the ideal people to help, advise and encourage coaches locally. They can help coaches not simply to progress to the next level, but in the overall quality of the work they are doing. Training mentors is a complex

task for the national governing bodies, and Champion Coaching has helped considerably.

Over 500 senior coaches have attended the workshops so far; here are some of their comments:

A very worthwhile four hours.

Gordon Lord/Head Coach, Bath

A most enjoyable enlightening day which has posed several interesting personal challenges.

Sean Hooper/Head Coach, Restormel

I did not feel originally I would gain much from this course ... How wrong I was!! It was super!!

Tony Corner/Head Coach, Bath

Master Coach (proposed equivalent S/NVQ level 4)

Most of these coaches, already highly qualified, were offered a custom package through the scholarship programme. Most coaches at this level also attended the Senior Coach workshop, as in some cases they were mentoring the mentors.

Supporting sport programmes

Helping children move from participation to performance is more than improving their skills and techniques. It also means helping them change their attitude and develop a sense of motivation and responsibility for their own progress. Right from the start of the pilot project it was felt that the children needed a way of recording their progress through the sport programme and planning their sporting future.

The *Sport Log* gives the children this opportunity, and encourages them to set targets with their coaches for improving their attitude to sport as well as developing specific skills and techniques. The sporting attitude section examines the children's attitudes to:

- Coaches
- Working with others
- Rules and decisions
- Learning by themselves
- Taking responsibility for their own progress.

Concerned coaches at the London Playing Fields Society programme

The Sports Councils acknowledge the problem of the gap between participation and performance levels, particularly for children. Champion Coaching has begun to help fill this gap by creating more 'child-friendly' opportunities for the performance-motivated youngster.

The partnership approach of Champion Coaching has started to facilitate several positive changes. These include:

● providing a framework for developments at performance level

● bringing together the agencies to focus on performance sport

● identifying the service available from governing bodies in local areas

● helping parents understand their role and the needs of children in sport

● identifying communities that are ready to offer performance sport.

Young people

Champion Coaching has demonstrated how young people respond so positively to high-quality sports coaching.

Peter Barson/Regional Director, South West Region

All the Sports Councils place great emphasis on their policies for young people. Champion Coaching's influence on youth sport development has spread far wider than the present forty-four schemes.

The 'readiness factors' schedule (see Chapter 2) that is distributed to all local authorities which express an interest in Champion Coaching is designed to help communities assess their existing provision for youth sport. The desire to be involved provokes much thought and activity within local authorities, and even if they are initially unsuccessful in their application to join Champion Coaching they may still integrate its philosophy and framework into their work with young people.

Most importantly, the Champion Coaching structure based on Youth Sport Managers and Youth Sport Advisory Groups is entirely complementary to the Sports Councils' policies that advocate local co-ordination and action.

Coach education and development

Through coach education the Youth Sport Manager and the district council have the potential to influence coaching structures and coach development.

Maureen Cusdin/Sports Council for Northern Ireland

The coach education programme in Champion Coaching is extensive and well balanced. This has led to an integration of the core knowledge and sport-specific elements that will be the basis of National Vocational Qualifications.

Coach education is an essential part of any coach development strategy, and the quality and format of the materials available in Champion Coaching has made it accessible to many. The structure of the coach education process also means that coaches are able to make logical progress from one level to another.

At the strategic level Champion Coaching has led to other advances. These include:

● the development of databases to record the details of coaches

● co-operation between neighboring schemes to promote coach development

● faster implementation of regional strategies

● recognition of the importance of formal and informal mentoring in coaching.

Community awareness of performance sport

Champion Coaching has brought the requirements of performance sport to the attention of the main providers, and has been readily accepted by all the networks currently involved in youth sport development.

George Jones/Regional Manager, Wales

Chapter 5

The national governing bodies

National governing bodies administer their sport while operating with workforces that are 90 per cent voluntary. In general they have limited resources with which to carry out the huge task of providing for the many and varied requirements of their sport at national and local level. However, the highly committed volunteer workers take an enormous pride in the job, and they care passionately about their sport and opportunities for young people.

The fifteen sports currently included in Champion Coaching are represented by nearly thirty governing bodies from England, Wales and Northern Ireland. Like local authorities, national governing bodies apply to be part of Champion Coaching, and have to fulfil certain criteria to be successful. These are to:

● provide a liaison officer

● set up an agreed scholarship programme at S/NVQ levels 2, 3 and 4

● attend regular development meetings to help position Champion Coaching within the governing-body structures

● agree to disseminate information on Champion Coaching

● have a commitment to fair play

● have a commitment to equal opportunities

● agree to clarify and check the qualifications of their coaches

● deliver their award schemes and courses locally.

The application system starts the partnership between the governing body and Champion Coaching. Once established, the partnership continues, with meetings between Champion Coaching and the governing body liaison officer three or four times a year. The purpose of these meetings is to:

● review their sport's structure for the development of players and coaches

● identify regional/county coaching co-ordinators to assist the schemes

● develop assessment tools for the selection of performance-motivated children

● help arrange national governing body clinics

● review the Champion Coaching Kit and to assist with improving it

● help to develop the scholarship process.

All the national governing bodies come to a joint meeting once a year where the multi-sport nature of Champion Coaching is revealed most powerfully.

It is not surprising that Champion Coaching has been seen as a real opportunity by most governing bodies to extend or consolidate their work in the following areas:

● partnerships with local authorities

● the local development of sport

● performance-motivated children

● exit routes

● recruiting coaches and coach development

● mentoring and assessing.

Partnerships with local authorities

Much more contact has been made with local authorities, and we are tapping into new people and new resources of which we were unaware before the Champion Coaching programme.

Betty Galsworthy/All-England Netball Association

Champion Coaching is proud of its association with all its partners, but an important development has been the increased dialogue between the partners themselves. The real power of partnership is measured by the links between *all* the partners and the way in which they co-operate.

Although national governing bodies have found that the needs of local authorities vary greatly, they have also realised that there are certain common threads between all the local authorities, of which the strongest is the desire to provide quality sporting opportunities for young people. As Youth Sport Managers have contacted national governing bodies for approval of coaches, and as the governing bodies have offered advice and helped schemes to provide award courses, clearer links have been established. The result is a much greater understanding between the local authorities and the national governing bodies.

Part of the strongly supported hockey programme in Humberside

The local development of sport

Champion Coaching has enabled our Development Officers to see how a partnership with local authorities is such a vital ingredient for our sport to develop. It has provided a model that can be repeated around the regions.

Don Parker/English Table Tennis Association

Not all governing bodies have a Development Officer network. Where they have, Champion Coaching has provided a useful model for the Development Officers in developing opportunities for children at the performance level; it has also increased the range of contacts for the Development Officers. Examples of these benefits include:

- Enabling National Cricket Association Development Officers to help local authorities find coaches and to give information about local clubs.

- The Football Association and its three Development Officers are working with Champion Coaching to promote football for girls and to recruit women coaches.

- All-England Netball Association Development Officers have a higher local profile through their involvement with Champion Coaching schemes.

- The Rugby Football League are using Champion Coaching as a model for future work beyond the participation level.

- The Squash Rackets Association are using Champion Coaching as the next step for their development project 'Squash in Schools'.

- The Rugby Football Union's Youth Development Officers have made additional links with schools they originally targeted with roadshows and beginner sessions.

Local development is also being assisted by an increased local profile for all the Champion Coaching sports.

Champion Coaching has enhanced Rugby League's profile among a variety of agencies. More people have become aware of what is on offer.

Tom O'Donovan/Rugby Football League

Enhancing the local profile is particularly important for the smaller national governing bodies. The important links that Champion Coaching has established between schools, facilities and clubs have enabled all governing bodies to benefit from greater publicity at the grass-roots level. Two examples are:

- The Welsh Hockey Association is now encouraging sport programmes in areas where hockey is a minority sport.

- Champion Coaching has helped make PE teachers and girls more aware of local opportunities in girls' football.

Performance-motivated children

The difficulties in co-ordinating opportunities for children is very challenging for the national governing bodies. At present, identifying and guiding the child who needs to progress is often left to individual coaches who may or may not know the next step. In many cases the pathway from participation to performance is unclear both to the coach and to the child.

Champion Coaching has enabled national governing bodies to think about their definition of the 'performance child' and also to start improving the performance pathways in their sport. The following are good examples:

- The All-England Netball Association has thought through the criteria for performance and made them available to YSMs and coaches.

- The Welsh Table Tennis Association has developed clearer routes from performance to excellence.

- The Hockey Association has included Champion Coaching in its four-year development plan.

Champion Coaching 1993

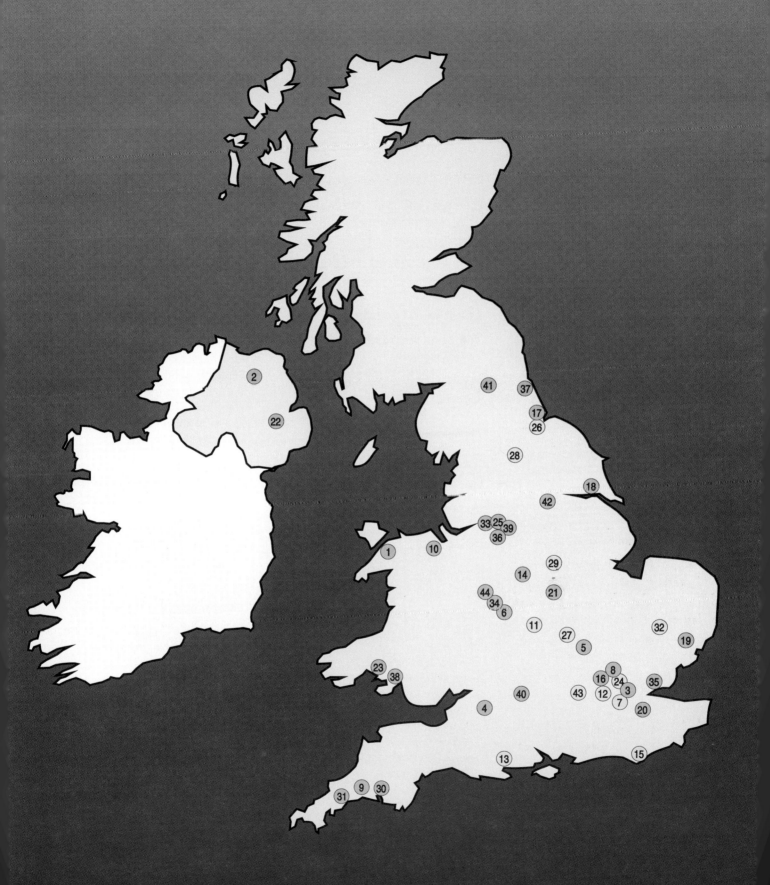

● Schemes dating from 1991

● New schemes in 1992/93

Hyfforddiant Penigamp Arfon Arfon Champion Coaching

1991 1992 1993 1994

Stand on top of Snowdon and look north-west towards Anglesey; the Borough of Arfon will be spread before you. Arfon includes the city of Bangor and Caernarfon town, and has some of the most striking scenery in the British Isles.

Champion Coaching is to be introduced here this September, with badminton, hockey, rugby (union, of course — this is Wales) and tennis as the first four sports.

Youth Sport Manager: Aled Roberts

Scheme address:
Arfon Borough Council
Penrallt
Caernarfon
Gwynedd LL55 1BN

Telephone: 0286 673113 ext. 505

Local management

The Youth Sport Manager is employed by Arfon Borough Council.

Time spent on scheme: **6 hours per week.**

Management structure

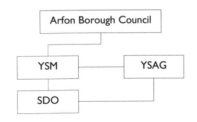

Youth Sport Advisory Group members
YSM
PE Adviser
Two PE teachers
Leisure Contract Manager (facilities)
Chairman, local Sports Council
NCF Development Officer
Sports Council for Wales Regional
  Officer
Sports Development Officer, Arfon
  Borough Council

Recruiting the children

All the children were nominated by their PE teachers. Schools were asked to nominate a set number of pupils, and all the children nominated were selected.

Contribution from the children £5.00 for the ten-week course.

Promoting the scheme

Schools were contacted through the local PE Adviser and their PE staff. The PE staff passed on information about the scheme to the children. Coaches were found through personal contacts and local knowledge. Press coverage was gained via the local papers.

Strengths of the scheme

● The creation of a new, local 'bank' of motivated coaches.

● The stronger links that have been made with PE teachers and schools.

Scope for development

We will be expanding the number of sports offered in years two and three.

The sport programmes

Sport	No. of sport progs	Sex	Ability level	Age group	No. children nomin'd	No. children selected	No. exit routes	No. new exit routes
Badminton	1	Mixed	Part	12–14	–	36	4	1
Hockey	1	Girls	Part	12–14	–	48	3	1
Rugby Union	2	Boys	Part	12–14	–	90	4	1
Tennis	1	Mixed	Part	12–14	–	36	3	1

* All sport programmes started in September.

The facilities

Sport	Facility	Owned by	Hourly rate
Badminton	Sports hall	Arfon Borough Council	£1.70 per crt
Hockey	Synthetic pitch	Arfon Borough Council	£10.00
Rugby Union	Club facilities	Local rugby clubs	Free
Tennis	Courts (school)	Gwynedd County Council	Free
	Indoor tennis centre	Arfon Borough Council	£4.00 per crt

The coaches

Sport	No. head coaches	No. coaches	Scholarship work at proposed equivalent S/NVQ level 2		Scholarship work at proposed equivalent S/NVQ level 3	
			NCF	NGB	NCF	NGB
Badminton	1	2	2	2	1	1
Hockey	1	4	4	4	1	1
Rugby Union	2	6	6	6	3	3
Tennis	1	2	2	2	1	–

Photo from Clwyd

New links

Existing links between the Sports Development Officers and the schools have been strengthened, and additional links have been made. More than fifty sports clubs have become involved with Champion Coaching.

Other comments

Birmingham's scheme is integrated into the city's 'Sports Development Strategy for the 90s'. The success of the scheme is largely due to the commitment, enthusiasm and professional expertise of our Sports Development Officers.

Our recipe for success is a cocktail involving many agencies. This first taste has been a good one.

by specific information about each sport. We relied on the teachers to tell the children about the scheme and to display posters. Once Champion Coaching was up and running, clubs were notified and invited to the coaching sessions. All coaches were approached personally by the mentor on each course and were then approved by the governing bodies.

Articles about the scheme appeared in the *Voice* council newspaper, which is distributed to over 200,000 Birmingham homes.

Strengths of the scheme

● By providing an opportunity for all fourteen sports to be part of the scheme, we ensured that the sports chosen were ready to take part, and that there were good exit routes.

● The setting up of sport-specific sub-groups to manage the individual sports meant that putting a sport 'on the ground' could be achieved effectively.

Scope for development

By looking at fourteen sports at the beginning of this year, and by setting up sub-groups, we have paved the way for the planned development of fifty-four sport programmes in our third year. This could not be managed by the YSM

and assistant YSMs, but only by a network of sport-specific sub-groups across the city.

The sport programmes

Sport	No. of sport progs	Sex	Ability level	Age group	No. children nomin'd	No. children selected	No. exit routes	No. new exit routes
Basketball	2	Mixed	Part	11–12	70	70	6	4
Cricket*	1	‡	‡	‡	†	‡	‡	‡
Hockey	2	Mixed	Part/perf	11–12	60	60	15	12
Netball	1	Girls	Part/perf	11–12	45	45	12	10
Squash*	2	‡	‡	‡	‡	‡	‡	‡
Swimming*	1	‡	‡	‡	‡	‡	‡	‡

* Sport programmes started in September.

The facilities

Sport	Facility	Owned by	Hourly rate
Basketball	Sports hall (school)	Grant-maintained school	£6.00
	Sports hall (leisure centre)	Birmingham City Council	£5.00
Cricket	Sports hall (leisure centre)	Birmingham City Council	£15.00
Hockey	Artificial pitch (school)	Birmingham City Council	£5.00
	Artificial pitch	Birmingham City Council	£12.50
Netball	Outdoor courts (dual use)	Birmingham City Council	£3.00
Squash	Courts (private)	Private club	Free
	Courts (college)	Birmingham City Council	Free
Swimming	Pool (leisure centre)	Birmingham City Council	£32.00

The coaches

Sport	No. head coaches	No. coaches	Scholarship work at proposed equivalent S/NVQ level 2 NCF	NGB	Scholarship work at proposed equivalent S/NVQ level 3 NCF	NGB
Basketball	2	4	1	–	5	–
Hockey	2	4	4	4	2	1
Netball	1	3	3	3	1	1

Bromley Champion Coaching

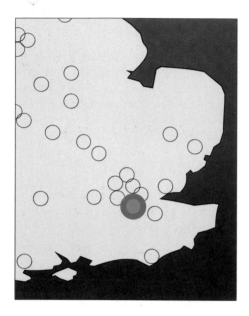

Bromley is the largest of the London Boroughs. It fills a large slice of the south-east of the metropolis and extends out to the fringes of Kent.

Bromley ran Champion Coaching in year one and set out to build on the lessons they learnt. This year the big difficulty was not transporting the children — it was finding them! Even though there was a well-established steering group in operation, and the Borough employed active development officers for the scheme, the response from the schools when invited to nominate children was very disappointing. Although this delayed part of the sport programme, things are now progressing well as this book goes to press, and by working with teachers Bromley are actively pursuing alternative ways of recruiting children.

Youth Sport Manager: Louise Aitken

Scheme address:
Sport Development Unit
Central Library
High Street
Bromley BR1 1EX

Telephone: 081-464 3333 ext. 3637

Local management

The Youth Sport Manager is employed by the London Borough of Bromley as Assistant Sports Development Officer.

Time spent on scheme: 1 day per week.

Youth Sport Advisory Group members
YSM
Assistant Sports Development Manager
Dual-use Development Officer
PE Curriculum Co-ordinator
Sports Council Regional Officer
NCF Development Officer
Chairman, Bromley Borough Hockey
  Development Scheme
Head coach
Teacher

Recruiting the children

PE teachers nominated children. Unfortunately, the response was very poor — many teachers felt that they did not have the time to take part in the nomination process being used.

Management structure

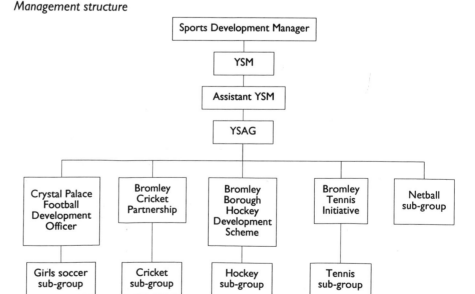

Scope for development

As yet, we have no plans for expanding/developing the scheme.

New links

As a result of the child selection process, schools are now in closer contact both with each other and the Leisure Department. Our links with clubs, governing bodies, the NCF and neighbouring district councils have also been strengthened.

The sport programmes

Sport	No. of sport progs	Sex	Ability level	Age group	No. children nomin'd	No. children selected	No. exit routes	No. new exit routes
Cricket*	1	Boys	Perf	12–14	30	30	‡	‡
Hockey	1	Girls	Perf	12–14	30	30	5	5
Netball	1	Girls	Perf	12–14	35	35	5	5
Rugby Union	1	Boys	Perf	12–14	45	45	5	2
Tennis*	2	Mixed	Perf	12–14	60	60	4	4

* Sport programmes started in September.

The facilities

Sport	Facility	Owned by	Hourly rate
Cricket	Facilities (club)	Callington Cricket Club	Free
Hockey	Pitch (school)	Cornwall County Council	Free
Netball	Court (school)	Cornwall County Council	Free
Rugby Union	Pitch (club)	Liskeard Rugby Club	Free
Tennis	Courts (school)	Cornwall County Council	Free

The coaches

Sport	No. head coaches	No. coaches	Scholarship work at proposed equivalent S/NVQ level 2		Scholarship work at proposed equivalent S/NVQ level 3	
			NCF	NGB	NCF	NGB
Hockey	1	2	2	2	–	–
Netball	1	2	1	1	2	1
Rugby Union	1	2	2	2	–	–
Tennis	1	2	2	2	1	1

Champion Coaching in Clwyd

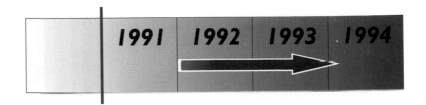

1991 | 1992 | 1993 | 1994

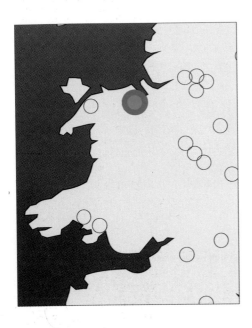

Champion Coaching in Clwyd is a county-wide scheme, rather than being confined to a town or city. The selection of seven sports is ambitious for a first year, and at times in the preliminary stages it looked as if it would be impossible to recruit sufficient coaches with the right qualities. Fortunately this problem was resolved in time for the full programme to run.

County YSM Shaun Hallett has worked closely with District Council colleagues in ensuring that Champion Coaching has a real local delivery. Champion Coaching has enabled much of the planning work carried out through the Clwyd Sportslink scheme to come alive.

See page 140 for a fuller profile of the Clwyd scheme.

Youth Sport Manager: Shaun Hallett

Scheme address:
Sportslink
Education Department
Shire Hall
Mold
Clwyd
CH7 6ND

Telephone: 0352 702747

Local management

The Youth Sport Manager is employed by Clwyd County Council as Community Sport and Recreation Manager.

Time spent on scheme: 10–15 hours per week.

Management structure

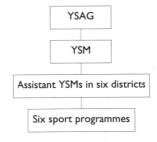

```
YSAG
  |
YSM
  |
Assistant YSMs in six districts
  |
Six sport programmes
```

Youth Sport Advisory Group members
YSM
Deputy Director, Clwyd County
    Council Education Department
PE Adviser/Inspector
Sports Council for Wales Regional
    Manager
Chief Tourism and Leisure Officer,
    Rhuddla Borough Council
Head of Leisure, Delyn Borough
    Council
Youth Service Officer, Clwyd County
    Council
Football Development Officer,
    Wrexham FC
Sports Development Officer, Alyn and
    Deeside District Council
NCF Development Officer
Head of PE
Sport Development Officer, Clwyd
    County Council

Recruiting the children

Children were pre-selected by teachers at each participating school. They all attended a 'meet the coaches' day, where some of them were diverted to other existing sport programmes.

Contribution from the children
£1.00 per session. Children who receive free school meals are exempt from payment.

Involving the parents

The parents organised a rota system for transporting children to and from the sport programme venues.

Promoting the scheme

Children were informed about the scheme by their PE teachers and the Champion Coaching posters displayed in schools. A regular newsletter was sent to schools and clubs. Coaches were contacted through governing bodies and local clubs.

Champion Coaching in Clwyd has been featured in ten newspaper articles, two local radio programmes and a regional television programme.

Strengths of the scheme

- We already had established links with schools and district councils via Sportslink Clwyd.

- Most of the facilities are owned by Clwyd County Council, and were available free of charge.

- The support of the governing bodies, who recognised the benefits of the scheme.

- The fact that all the agencies involved in the scheme are working together rather than individually.

Scope for development

The scheme could be developed both by increasing the number of sports and by recruiting more children to the sport programmes. To do this, we need to utilise the newly qualified coaches to manage the sessions, and also give district councils more responsibility.

New links

We have strengthened the links already in place through Sportslink Clwyd, and have established a mechanism for attracting children into well-structured clubs.

Other comments

Because of the size of the scheme, the consultation process was perhaps too long for a fast and effective start to the sport programmes.

The quality of the coaches has underpinned the success of the scheme, and I am confident that the coaching delivery to the children has been first class.

The coaching sessions took place immediately after school, largely utilising school facilities. The children thus learned to recognise their school as a community resource.

The sport programmes

Sport	No. of sport progs	Sex	Ability level	Age group	No. children nomin'd	No. children selected	No. exit routes	No. new exit routes
Badminton*	3	Mixed	Perf	11–14	‡	134	5	1
Hockey	2	Mixed	Perf	11–14	‡	50	2	–
Netball	3	Girls	Perf	11–14	‡	150	4	1
Rugby Union	3	Boys	Perf	11–14	‡	100	11	–
Soccer	2	Girls	Part	11–14	‡	110	5	4
	1	Boys	Perf	11–14	‡	30	1	–
Tennis	2	Mixed	Perf	11–14	‡	50	3	–

* Sport programme started in September.

The facilities

Sport	Facility	Owned by	Hourly rate
Hockey	Pitches (schools)	Clwyd County Council	Free
Netball	Sports hall (leisure centre)	Colwyn Borough Council	Free
Rugby Union	Pitches (schools)	Clwyd County Council	Free
	Pitch (leisure centre)	Colwyn Borough Council	Free
Soccer	Pitch (school)	Clwyd County Council	Free
	Pitch (dual use sports ctr)	Clwyd CC/Delyn BC	Free
	Pitch	Howarden Community Council	Free
Tennis	Courts (schools)	Clwyd County Council	Free

The coaches

Sport		No. head coaches	No. coaches	Scholarship work at proposed equivalent S/NVQ level 2		Scholarship work at proposed equivalent S/NVQ level 3	
				NCF	NGB	NCF	NGB
Badminton		3	6	6	6	3	1
Hockey		2	5	3	2	3	–
Netball		3	9	9	9	2	–
Rugby		3	8	3	3	6	2
Soccer	(girls)	2	7	7	4	4	2
	(boys)	1	3	3	–	1	–
Tennis		2	5	5	5	2	1

Coventry Sports Training Programme in association with Champion Coaching

1991 | 1992 | 1993 | 1994

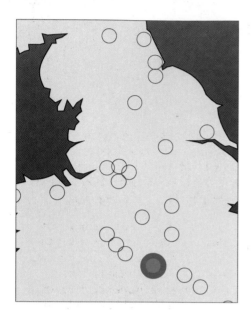

Coventry was in Champion Coaching from the beginning. In their first year they had learned that a lot of problems would be avoided if children, clubs and facilities were kept quite close to each other. So this time quite a lot of planning went into identifying 'clusters' of these in different catchment areas across the city.

There has been a big investment of time in building close relationships with the schools, which has resulted in Coventry enjoying free access to school facilities — a feature which will make them the envy of many other schemes.

A particular strength of Champion Coaching which YSM Fiona Fortune pinpoints is that the participating sports feel a degree of 'ownership' of it. They see the benefits and want the whole programme to expand.

Youth Sport Manager: Fiona Fortune

Scheme address:
Leisure Services Department
Coventry City Council
Floor 5, Spire House
New Union Street
Coventry
CV1 2PS

Telephone: 0203 832446

Local management

The Youth Sport Manager is employed by Coventry City Council Leisure Services Department.

Time spent on scheme: 12–15 hours per week.

Management structure

Youth Sport Advisory Group members
No YSAG operates in the Coventry scheme.

Recruiting the children

Children who met the criteria laid down to the schools were allowed to apply for the sport programmes.

Assessment sessions were held for sport programmes in netball and basketball. All the children who attended were of the right standard, and so all were accommodated.

Contribution from the children
£1.00 per session (50p for children holding a Passport to Leisure and Learning).

Involving the parents

Head coaches were encouraged to involve those parents who attended their children's sessions. Parents will be invited to the netball and basketball tournaments which are to be run, where they will be able to meet the people involved in running the scheme.

Promoting the scheme

Children were informed about the scheme through Champion Coaching posters and leaflets. Some of the schools knew about the scheme, having taken part last year. Others were informed at a meeting for heads of PE. Potential clubs were targeted with leaflets and contacted individually. Coaches were contacted individually or through key people in the relevant sports.

We are planning to produce a newsletter to coaches and PE teachers, and will feed information to community college newsletters. The tournaments are to be publicised in the local press.

Strengths of the scheme

- The political support we have allows for an easier working relationship with other departments and external organisations.

- The fact that some of the scheme administration was taken on by other agencies meant that more time was available for other initiatives and the chances of continuing the scheme were increased.

- The sports themselves were keen to support the scheme and see it succeed because they could see its benefits.

- Champion Coaching has a clear role and is easy to identify within the existing sports structure.

- Our 'cluster approach' minimises the transport needed for the children. This is a lesson we learnt from last year.

- Good links with schools enabled us to gain use of facilities free of charge.

Scope for development

The scheme could be expanded to cover the whole of the city (at present the scheme covers only the north), and could include extra sports, either city-wide or in a cluster. The other possible area for expansion would be into other age groups.

New links

Secondary-school links have been strengthened, and new links developed at three primary schools for swimming. We have improved contacts with the City of Coventry Swimming Club, Coventry Crusaders and the RFU Youth Development Officer, and have forged new links with two netball clubs.

Other comments

Key people have come forward and have ensured that the courses have been successful. Many have acted as head coaches and also looked after the administration of the programmes.

The teachers' commitment has been tremendous and many have acted as coaches. All the coaches have done an amazing job — it is obvious that the children have developed their sport skills, social skills and sense of fair play.

The sport programmes

Sport	No. of sport progs	Sex	Ability level	Age group	No. children nomin'd	No. children selected	No. exit routes	No. new exit routes
Basketball	3	Mixed	Perf	12–14	65	65	1	1
Netball	2	Girls	Perf	12–14	50	50	5	1
Rugby Union	1	Boys	Perf	13–14	16	16	–	–
Swimming	1	Mixed	Perf	11–12	24	24	*	1

* Exit route via the City of Coventry Swimming Club squad programme.
More sport programmes commence in October.

The facilities

Sport	Facility	Owned by	Hourly rate
Basketball	Sports halls (schools)	Coventry City Council	Free
Netball	Court (sports centre)	Coventry City Council	Free
	Court (school)	Coventry City Council	Free
Rugby Union	Pitch (school)	Coventry City Council	Free
Swimming	Pool (school)	Coventry City Council	Free

The coaches

Sport	No. head coaches	No. coaches	Scholarship work at proposed equivalent S/NVQ level 2 NCF	Scholarship work at proposed equivalent S/NVQ level 2 NGB	Scholarship work at proposed equivalent S/NVQ level 3 NCF	Scholarship work at proposed equivalent S/NVQ level 3 NGB
Basketball	3	6	3	1	3	3
Netball	2	5	4	4	3	1
Rugby Union	1	2	–	–	3	3
Swimming	1	2	–	–	3	3

Ealing Champion Coaching

1991 | 1992 | 1993 | 1994

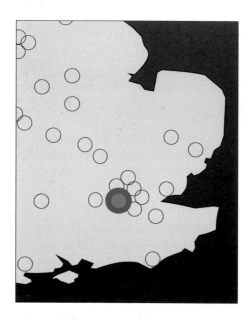

Ealing is a West-London borough with a very mixed population. It ran one of the first Champion Coaching schemes and expanded it this year. New sports for 1993 are badminton, cricket and volleyball — the latter being helped by the prominent Polonia volleyball club being in the area.

Links with schools in the Borough are not as strong as they could be, and Champion Coaching is playing a part in strengthening these. Similarly, parental support has been less than in other parts of the country, but it is hoped that implementation of Young People in Sport workshops will assist this.

Youth Sport Manager: Sarah Wisely

Scheme address:
Sports Development
London Borough of Ealing
3rd Floor
Perceval House
14–16 Uxbridge Road
Ealing
W5 2HL

Telephone: 081-758 5903

Local management

The Youth Sport Manager is employed by the London Borough of Ealing as Sports Development Officer (Performance).

Time spent on scheme: Difficult to estimate, but probably 1½–2 days per week.

Management structure

Youth Sport Advisory Group members
YSM
Head of PE
Sports Council Regional/Liaison Officer
RFU Youth Development Officer
Borough Recreation Manager
Chairman, Polonia Volleyball Club
Ex-PE Adviser
NCF Development Officer

Recruiting the children

Nomination forms were distributed to all the relevant schools within the borough. Assessment sessions were than held for each sport.

Contribution from the children
£12.00 per eight-week programme (sessions lasted 1½ hours).

Promoting the scheme

Children were informed about the scheme with newsletters to their school, press releases and school visits. Meetings were held with PE teachers. Clubs were contacted through the development officers and governing bodies of the relevant sports. Governing bodies were also used to find coaches.

There were numerous press releases in the local papers which outlined the aims and objectives of the scheme.

Involving the parents

The parents did not really get involved, but they have been complimentary and supportive.

Strengths of the scheme

● The coaches were all committed to coaching and are keen to be involved.

● The children display great enthusiasm and enjoyment. They are grateful for their opportunity, and there have been very few absentees.

Scope for development

There is definitely room to develop better relationships with schools in the Borough. This could be achieved through support from the education department, together with a new approach to the teachers' involvement in the scheme.

New links

We are building links with junior schools (although this is limited because of reorganisation). We are also working closely with rugby union, badminton and volleyball clubs.

Other comments

The involvement of the schools in the nomination process has been a major problem, but the children are finding that Champion Coaching is valuable for their sporting development.

Operating within the local-authority sports-development budget has meant that participant fees have had to cover the costs of coaches and facilities. It has also meant that Champion Coaching has not had such a high priority as other schemes.

Running a scheme could be a full-time job, and I only wish I could devote more time to it. Hopefully the scheme will consolidate over the coming years.

The sport programmes

Sport	No. of sport progs	Sex	Ability level	Age group	No. children nomin'd	No. children selected	No. exit routes	No. new exit routes
Badminton*	‡	‡	‡	‡	‡	‡	‡	‡
Basketball*	‡	‡	‡	‡	‡	‡	‡	‡
Cricket	1	Boys	Part/perf	11–12	39	31	4	–
Hockey*	‡	‡	‡	‡	‡	‡	‡	‡
Netball	1	Girls	Part/perf	11–12	53	30	2	–
Rugby Union*	‡	‡	‡	‡	‡	‡	‡	‡
Volleyball	1	Mixed	Part	11–12	31	24	1	–

* Sport programmes started in September.

The facilities

Sport	Facility	Owned by	Hourly rate
Cricket	Sports ground (club)	Barclays Bank	£50 for whole programme
Netball	Courts (dual use)	London Borough Ealing Ed. & Leis.	Free
Rugby Union	Pitch (club)	London Borough Ealing (lease)	Free
Volleyball	Sports hall (sports centre)	London Borough Ealing Ed. & Leis.	Free

The coaches

Sport	No. head coaches	No. coaches	Scholarship work at proposed equivalent S/NVQ level 2 NCF	NGB	Scholarship work at proposed equivalent S/NVQ level 3 NCF	NGB
Cricket	2	2	1	1	1	1
Netball	1	2	2	2	–	–
Volleyball	1	3	2	2	–	–

Strengths of the scheme

- The scheme provides a natural progression from existing participation schemes.

- The scholarships improve the quality of coaching and give coaches on other schemes something to aim for.

- The partnership philosophy, which brings together people from various agencies in a policy-making forum.

- The development of a co-ordinated coach-education programme across the West Midlands, allowing the sharing of ideas.

- The national prestige of the scheme filters down and is good for local PR.

- The Champion Coaching resources (*Sport Log*, workshop materials, Evans vouchers, etc.) make the scheme easier to promote by having something to bargain with.

Scope for development

Where coaches and exit routes permit, the East Staffordshire scheme could be expanded both by increasing the number of sport programmes for each sport and by expanding the range of sports. Both of these options would require additional funding which is not currently available.

New links

New school links have been made and those links already in existence have been strengthened. We are trying to form a junior section at an existing basketball club, and a completely new junior girls soccer club.

Other comments

Champion Coaching has helped us to develop and strengthen links with local schools, clubs, governing bodies and coaches, and is an important and integral part of our sports provision for the community.

The sport programmes

Sport	No. of sport progs	Sex	Ability level	Age group	No. children nomin'd	No. children selected	No. exit routes	No. new exit routes
Basketball*	1	Mixed	Perf	12–14	‡	Target 24	3	1
Hockey	1	Mixed	Perf	12–14	38	38	2	1
Netball*	2	Girls	Part	12–14	‡	Target 30	2	2
Soccer	1	Girls	Part	11–14	16	16	1	1
Table tennis*	1	Mixed	Part	11–14	‡	Target 20	2	1

* Sport programmes started in September.

The facilities

Sport	Facility	Owned by	Hourly rate
Basketball	Sports hall (leisure centre)	Borough of East Staffordshire	£17.87
Hockey	Pitch (leisure centre)	Borough of East Staffordshire	Free
Netball	Courts (club)	Ind Coope	Free
Soccer	Pitch (leisure centre)	Borough of East Staffordshire	£7.66
Table tennis	Table-tennis club	Staffordshire County Council	£8.00

The coaches

Sport	No. head coaches	No. coaches	Scholarship work at proposed equivalent S/NVQ level 2 NCF	NGB	Scholarship work at proposed equivalent S/NVQ level 3 NCF	NGB
Basketball	1	3	2	2	1	–
Hockey	1	2	2	2	1	1
Netball	1	3	2	2	1	1
Soccer	2	1	1	1	2	1
Table tennis	1	2	2	2	1	1

ACCESS
(Active Champion Coaching for East Sussex Schools)

1991 | 1992 | 1993 | 1994

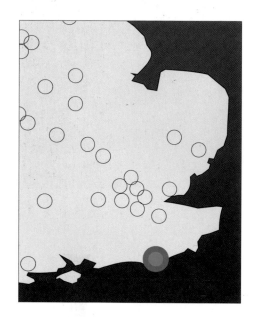

Youth Sport Manager: Robert Lake

Scheme address:
Chelsea School of PE
Hillbrow Cottage
Gaudick Road
Eastbourne
East Sussex
BN20 7SP

Telephone: 0273 643761

Local management

The Youth Sport Manager is employed by East Sussex County Council Education Department as County Sports Development Officer.

Time spent on scheme: 12 hours per week.

Management structure

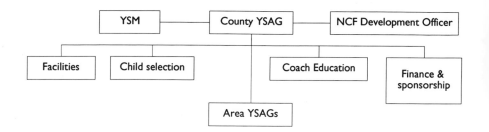

East Sussex is a large and populous area which includes Brighton, Eastbourne and Hastings. The Champion Coaching scheme is ambitious, and is planned to allow broader development. A particular feature is to take a very active lead in setting up exit routes where none existed previously. Even before the programme was half-complete, eight new junior sections in sports clubs were established, as well as two totally new clubs, complete with adult sections!

The structure of the East Sussex scheme is necessarily complex, with six additional YSAG Sub Groups loking at specific components of the Champion Coaching Blueprint. Co-operation between Champion Coaching and the schools is strong, and parental involvement is high: several new coaches have been recruited from this source.

County YSAG members
YSM
Headteacher
Head of PE
Head of PE/basketball head coach
Two teachers/basketball head coaches
South East Regional Netball
  Development Officer
South East Regional Table Tennis
  Development Officer

Sports Development Officers, Brighton
  Borough Council, Hove Borough
  Council, Eastbourne Borough Council
Youth Development Officer, Sussex
  County Cricket Club
Sports Council Regional Officer
NCF Development Officer
County PE Adviser
Chief Leisure Officer, Hove Borough
  Council

Recruiting the children

A letter and questionnaire were given by PE staff to children who they thought had the right ability and attitude for the scheme. Suitability was finally decided at sessions prior to the sport programmes.

Contribution from the children £1.00 per session.

Involving the parents

In a number of the sport programmes we have attracted parents as new coaches.

Promoting the scheme

Children were informed about the scheme through posters and school visits. Two of the sport programmes organised demonstrations to arouse interest. Personal visits were made to PE teachers in every school, headteachers were written to, and information appeared in the county publication *Education Scene*. Clubs and coaches were found through sport-specific development officers, county governing bodies and the local sports council.

Articles and adverts about Champion Coaching appeared in the local press.

Strengths of the scheme

- The partnership between local authorities and Champion Coaching.

- The support and guidance received from the NCF Development Officer.

- The use of area YSAGs.

- The support of schools, both in terms of the involvement of PE teachers and the willingness of headteachers to offer facilities free of charge in many cases.

- The coaches were carefully chosen and have been willing to take on a considerable workload.

- Facility holders were prepared to provide concessionary rates.

Scope for development

The local advisory groups could take on more responsibility for the day-to-day management of the scheme. The scheme could also benefit from greater involvement of the county and regional governing bodies. We plan to add five more sports in year two.

New links

New junior club sections have been established in all eight sports which will be part of the scheme, and we have set up two totally new netball clubs, which also have adult sections.

The sport programmes

Sport	No. of sport progs	Sex	Ability level	Age group	No. children nomin'd	No. children selected	No. exit routes	No. new exit routes
Basketball	2	Boys/ mixed	Part/perf	13–14	60	60	2	2
Cricket	3	Boys	Part/perf	12–14	100	65	3	3
Netball	1	Girls	Part/perf	13–14	28	28	1	1
Table Tennis	2	Mixed	Part/perf	12–14	40	28	2	2

The facilities

Sport	Facility	Owned by	Hourly rate
Basketball	Sports hall (leisure centre)	Brighton Borough Council	£4.90 session
	Sports hall (school)	East Sussex County Council	£7.50
Cricket	Pitches (school)	East Sussex County Council	Free
	Pitches (community coll.)	East Sussex County Council	Free
Netball	Court (dual use)	Walden District Council/	£5.00
		East Sussex County Council	(first hour free)
Table tennis	Sports hall (school)	East Sussex County Council	Free
	Sports hall (commun. coll.)	East Sussex County Council	Free

The coaches

Sport	No. head coaches	No. coaches	Scholarship work at proposed equivalent S/NVQ level 2 NCF	Scholarship work at proposed equivalent S/NVQ level 2 NGB	Scholarship work at proposed equivalent S/NVQ level 3 NCF	Scholarship work at proposed equivalent S/NVQ level 3 NGB
Basketball	2	7	5	5	2	2
Cricket	2	9	6	6	5	3
Netball	1	2	2	2	1	1
Table Tennis	2	3	1	1	3	3

Strengths of the scheme

● The scheme opens new opportunities for the 11–16 age group, especially for girls, through local venues and good coaching.

Scope for development

The scheme could be developed in many ways — by including more sports, by expanding into county schools, and encouraging more nominations. We also aim to develop cluster groups of schools to feed programmes at local venues.

New links

We have strengthened links with schools, especially through the dual use of their facilities. We have also made better links with governing bodies, which has helped us to find more coaches.

The sport programmes

Sport	No. of sport progs	Sex	Ability level	Age group	No. children nomin'd	No. children selected	No. exit routes	No. new exit routes
Hockey	1	Mixed	Part	12–13	15	15	‡	‡
Netball	2	Girls	Perf	12–14	40	40	‡	‡
Rugby Union	1	Boys	Perf	13	30	30	‡	‡
Soccer	1	Girls	Perf	12–14	15	15	‡	‡
Tennis	1	Mixed	Part	12	30	30	‡	‡

The facilities

Sport	Facility	Owned by	Hourly rate
Hockey	Pitch (sports centre)	Leicester City Council	£7.50
Netball	Netball centre	Leicester City Council	£7.50
	Court (school)	Leicestershire Education Authority	£7.50
Rugby Union	Facilities (club)	Ayestonians Rugby Club	Free
Soccer	Sports centre	Leicester City Council	£11.00
Tennis	Courts (leisure centre)	Leicester City Council	£11.00

The coaches

Sport	No. head coaches	No. coaches	Scholarship work at proposed equivalent S/NVQ level 2 NCF	NGB	Scholarship work at proposed equivalent S/NVQ level 3 NCF	NGB
Hockey	1	–	2	–	1	–
Netball	2	6	4	–	2	–
Rugby Union	1	3	2	–	1	–
Soccer	1	3	–	–	3	–
Tennis	1	–	2	–	1	–

Lisburn Champion Coaching

1991 1992 1993 1994

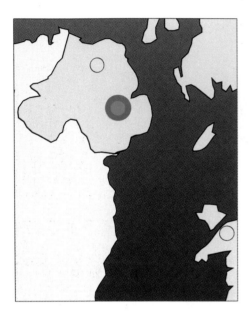

Lisburn is a few miles south of Belfast. The scheme does not start to run until the end of September, when badminton, football, hockey, netball and swimming get under way.

In Northern Ireland it has been customary for the various agencies involved in youth sport to 'do their own thing' in isolation, and the YSM reports that Champion Coaching is prompting a previously unknown degree of co-operation.

Youth Sport Manager: Paul Whitten

Scheme address:
Borough Offices
The Square
Hillsborough
County Down
BT26 6AH

Telephone: 0846 682477

Local management

The Youth Sport Manager is employed by Lisburn Borough Council as Sports Development Officer.

Time spent on scheme: 2–2½ days per week.

Management structure

Lisburn Borough Council

Leisure Services Department (YSM)

Six sport programmes

Youth Sport Advisory Group members
YSM
Creative and Expressive Studies Adviser
Sports Council for Northern Ireland Officer
Divisional Youth Officer
Headteacher
PE teacher
Head coach
Facility manager

Other co-operating agencies
South Eastern Education and Library Board
Youth Service
Lisburn Sports Advisory Council

Recruiting the children

All the children were nominated by their teachers. Some who were already on coaching courses may have been recommended by a coach, but their nominations were still 'passed' by their teachers.

A two-hour assessment session was held in each sport at which the head coach and PE staff made the selections.

Contribution from the children
£10.00 per ten-week programme.

Promoting the scheme

Schools were informed about the scheme at teacher briefings and by personal visits. Children were informed by their teachers at school or by coaches at the 'try it' courses. Clubs were contacted by the YSM , by coaches and by letter. The coaches themselves were found through governing bodies and contacts within the sports.

We held a local and province-wide launch with the Sports Council for Northern Ireland. We also propose to hold parents' briefings and club briefings.

Strengths of the scheme

- The close working relationship with local clubs and schools.

- The cohesive programme of development and the effective use of resources means less duplication and better communication.

- The formation of a pool of high-quality coaches who are trusted by teachers.

- Champion Coaching has provided children with a means of progression from existing, more static participation schemes.

Scope for development

To an extent, this has been an experimental year. Years two and three will see real development.

New links

Overall, there has been greater contact with various agencies, in particular the South Eastern Education and Library Board and the Creative and Expressive Studies Adviser. Hopefully, better links will develop with schools and clubs in future scheme years.

Other comments

Comments from all the agencies involved in the scheme stress the common-sense approach, which avoids duplication. We have several clubs and national governing bodies wishing to join next year.

The sport programmes

Sport	No. of sport progs	Sex	Ability level	Age group	No. children nomin'd	No. children selected	No. exit routes	No. new exit routes
Badminton	1	Mixed	Perf	11–14	‡	24	‡	‡
Hockey	1	Boys	Perf	11–14	‡	30	‡	‡
	1	Girls	Perf	11–14	‡	24	‡	‡
Netball	1	Girls	Perf	11–14	‡	24	‡	‡
Soccer	1	Boys	Perf	11–14	‡	24	‡	‡
Swimming	1	Mixed	Perf	11–14	‡	24	‡	‡

All sport programmes started in September.

The facilities

Sport	Facility	Owned by	Hourly rate
Badminton	Sports hall (leisure centre)	Lisburn Borough Council	£13.00
Hockey	Pitch (club)	Club	£15.00
Netball	Sports hall (school)	School	£13.00
Soccer	Floodlit pitch (leisure cntr)	Lisburn Borough Council	£13.75
Swimming	Pool (leisure centre)	Lisburn Borough Council	‡

The coaches

Sport	No. head coaches	No. coaches	Scholarship work at proposed equivalent S/NVQ level 2 NCF	NGB	Scholarship work at proposed equivalent S/NVQ level 3 NCF	NGB
Badminton	1	4	4	4	1	1
Hockey (boys)	1	3	3	3	1	–
(girls)	1	3	3	3	1	–
Netball	1	3	3	3	1	1
Soccer	1	3	3	2	–	–
Swimming	1	3	2	2	2	1

Recruiting the children

Children were usually recruited by being proposed by their schools. Proposals were by word of mouth from advertising and by request from parents.

If the programmes were over subscribed, coaches selected children from the nominees.

Contribution from the children £1.50 per session.

Promoting the scheme

Contact with schools was made through PE teachers, who told the children about the scheme. Clubs were contacted personally. Coaches were found through coaching secretaries and senior and county committees.

General publicity took the form of newsletters and posters.

Strengths of the scheme

● The fact that the scheme has a local sports development programme base ensures that it is not just a 'tacked on' ten weeks of activity.

● The scheme increases the opportunity for more children to receive high-quality coaching.

Scope for development

We could develop the scheme by more concentrated promotion within the districts — by tailoring existing opportunities and providing new ones where necessary. We should also ensure that the programmes link in with school competitions so that children can put into practice what they have learnt.

New links

Some schools have become centres for local sport — a magnet for school children in the area. We have found club coaches who are intent on establishing junior and youth sections. Contact has also been made with individual sport-development groups.

The sport programmes

Sport	No. of sport progs	Sex	Ability level	Age group	No. children nomin'd	No. children selected	No. exit routes	No. new exit routes
Basketball	3	Boys	Perf	12–14	100	100	2	1
Hockey	4	Mixed	Perf	12–14	120	90	4	1
Table tennis	1	Mixed	Perf	12–14	21	19	2	1
Tennis	3	Mixed	Part/perf	12–14	30 or 50	30 or 50	2	1
Volleyball*	‡	‡	‡	‡	‡	‡	‡	‡

* Sport programme started in September.

The facilities

Sport	Facility	Owned by	Hourly rate
Basketball	Youth centre	Daventry District Council	£10.00
	Sports hall (tech. coll.)	College	£15.00
Hockey	Artificial pitch	Northampton Borough Council	Free
	Artificial pitch (school)	Northamptonshire County Council	Free
Table tennis	Hall (school)	Northamptonshire County Council	‡
Tennis	Courts (school)	Northamptonshire County Council	£2.00 per crt
	Courts (school)	Northamptonshire County Council	Free

The coaches

Sport	No. head coaches	No. coaches	Scholarship work at proposed equivalent S/NVQ level 2 NCF	NGB	Scholarship work at proposed equivalent S/NVQ level 3 NCF	NGB
Basketball	3	5	2	1	3	–
Hockey	4	6	6	6	4	1
Table tennis	1	3	2	2	1	–
Tennis	3	6	3	3	3	–

North Yorkshire Coaching Forum — Champion Coaching

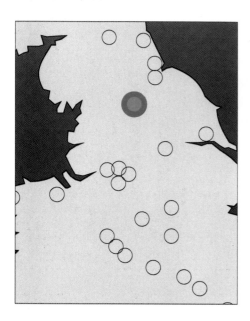

North Yorkshire is a very large county with a generally rural nature. In the first year of Champion Coaching only its Ryedale District was involved, but this time it has been joined by another five districts, under the county-wide banner of the North Yorkshire Coaching Forum.

Each of the districts has its own Youth Sport Advisory Group, and a particular feature of these is the involvement of parents. Some of the more remote districts had virtually no tradition of coach education, so the scheme has been an important catalyst in promoting this and in releasing otherwise-inaccessible funds to help school-age sport.

Youth Sport Manager: Chris Fielden

Scheme address:
Community Education
County Hall
Northallerton
North Yorkshire
DL7 8AE

Telephone: 0609 780780 ext. 2254

Local management

The Youth Sport Manager is employed by North Yorkshire County Council as Sports Development Officer.

Time spent on scheme: 32 hours per week on average.

Management structure

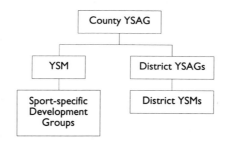

Youth Sport Advisory Group members
YSM
Sports/Recreation Officer from each
  district
Sports Council Liaison Officer
NCF Development Officer
Senior PE Adviser
Community Education Officer
School governor
Headteacher
Federation of Disability Sports
  Organisations representative
Coaches will be included after the
  coaches forum is established.

Other co-operating agencies
District councils
Community Education Service

Recruiting the children

The nomination and selection process varied from district to district. However, in general, PE teachers, voluntary sports clubs and the community education service were involved in nominations. Selection was done by coaches, teachers and club representatives at 'challenge days', competitions days, and other selection days. In some cases, all the nominated children were able to join the scheme.

Contribution from the children
Varied from district to district and from sport to sport. Ranged from £5 to £35 for the ten-week course.

Involving the parents

Some parents sat on the local YSAGs and some acted as assistant coaches. In certain areas they took part in family 'fun days' run to attract interest.

Promoting the scheme

Schools were contacted via their PE teachers and Youth officers. Schools were also sent leaflets and given presentations which the children attended. Children were also made aware of the scheme with posters and articles in the local press. Clubs were contacted through the district sports councils, were given presentations, and were invited to attend the challenge and fun days. Coaches were contacted via local sports councils and sport specific development groups. Their interest was sought in a questionnaire as part of the development of a local coaching strategy.

The value of the scheme was emphasised in local radio presentations, press coverage, and at meetings with members of district councils, local sports councils and sport-specific development groups.

Strengths of the scheme

● The co-ordinated approach, which not only provides opportunities for young people, but also forms partnerships and opens dialogues with other agencies.

● The scheme forms a permanent part of sport-specific development plans.

● The scheme is creating a coaching force which is competent to work with young people.

● Because of the way it has been marketed, the scheme 'belongs' to the community.

● The establishment of a 'coaches forum', where wider coaching issues can be addressed.

Scope for development

The scheme could be taken into new areas of the districts. It could also be expanded into new sports and provided with new exit routes.

New links

New links have been established and existing links strengthened between the district councils and schools. Some new junior clubs sections have been started.

The sport programmes

Sport	No. of sport progs	Sex	Ability level	Age group	No. children nomin'd	No. children selected	No. exit routes	No. new exit routes
Badminton	1	Mixed	Perf	12–14	50	20	12	1
Basketball*	2	Mixed	Part/perf	11–14	‡	120	5	5
Cricket	1	Mixed	Part	11–14	‡	16	–	–
Hockey	1	Mixed	Perf	11–14	103	68	2	1
Netball	1	Girls	Perf	11–14	85	56	2	1
	2	Girls	Part/perf	11–14	‡	90	1	1
Orienteering*	1	Mixed	Perf	12–14	50	20	12	5
	1	Mixed	Perf	11–14	‡	32	2	1
Rugby League*	1	Mixed	Perf	11	‡	30	–	–
	2	Mixed	Part/perf	11–14	‡	120	3	3
Rugby Union	1	Mixed	Part/perf	11–14	‡	60	1	1
Soccer*	1	Girls	Part/perf	11–14	‡	60	1	1
Swimming*	2	Mixed	Perf	11–14	‡	90	2	1
Table tennis	1	Mixed	Perf	11–14	28	26	–	–
	1	Mixed	Part/perf	11–14	‡	60	3	2

* Sport programmes started in September.

The facilities

Sport	Facility	Owned by	Hourly rate
Badminton	Sports hall (leisure centre)	Hambleton District Council	Free*
Basketball	Sports hall (dual use)	Selby DC/North Yorkshire CC	Free
	Gym/outside area	North Yorkshire County Council	Free
Cricket	Sports hall/field	Ryedale DC/North Yorkshire CC	£25.00
Hockey			
Netball	Sports hall	Ryedale DC/North Yorkshire CC	£25.00
	Sports hall/courts	North Yorkshire County Council	Free
Orienteering	Leisure centre	Hambleton District Council	Free*
Rugby League	Pitch	North Yorkshire County Council	Free
	Pitch (club)	Hewath ARLFC	Free
	Sports hall/outside	North Yorkshire County Council	Free
Rugby Union	Pitch (club)	Private club	Free
Soccer	Pitches (school/club)	North Yorkshire CC/private club	Free
Swimming			
Table tennis	Gym (leisure centre)	North Yorkshire County Council	Free
	Sports hall (sports centre)	Scarborough Borough Council	£8.40

*These facilities were subsidised; other free facilities were negotiated.

The coaches

Sport	No. head coaches	No. coaches	Scholarship work at proposed equivalent S/NVQ level 2 NCF	Scholarship work at proposed equivalent S/NVQ level 2 NGB	Scholarship work at proposed equivalent S/NVQ level 3 NCF	Scholarship work at proposed equivalent S/NVQ level 3 NGB
Badminton	1	1	1	1	1	1
Basketball	2	5	5	5	1	–
Cricket	1	1	–	–	–	1
Hockey	1	2	–	–	3	3
Netball	3	5	4	4	4	3
Orienteering	2	3	–	–	2	3
Rugby League	3	6	6	6	3	1
Rugby Union	1	2	2	2	2	1
Soccer	1	2	2	2	1	–
Swimming	1	6	6	6	1	1
Table tennis	2	7	4	4	2	2

Champion Coaching with the Sports Training Scheme (Nottinghamshire)

1991 1992 1993 1994

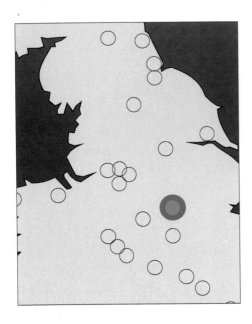

Nottinghamshire has a good history of youth sport development, so Champion Coaching has fertile soil in which to grow. There are many active coaches in the area who have seen the scheme as an opportunity to enhance their own education, and the NCF scholarships are being taken up with enthusiasm.

As an object lesson in 'catching them young', assistant coaches are being recruited from fifth-formers in some schools. These are all pupils who are taking the GCSE in physical education, and the knowledge gained from their NCF courses is helping them with preparation for their exams.

Naturally, as this is a county scheme it has many programmes running. Not all of these are 'easy' ones — for example, orienteering and squash are well represented. The venues are taken from a broad mixture of local education, district council and private club sources.

Youth Sport Manager: Sue Conner

Scheme address:
Nottinghamshire Sports Coaching Centre
Sycamore Complex
Hungerhill Road
St Anns
Nottingham

Telephone: 0602 624040

Local management

The Youth Sport Manager is employed by Nottinghamshire County Council Leisure Services.

Time spent on scheme: 12 hours per week.

Management structure

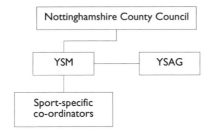

Youth Sport Advisory Group members
YSM
PE Inspector
Headteacher
PE teacher
NCF Development Officer
Local coach
Principal Officer (Sport)
PENS Project Officer
Sports Council Regional Officer
Chairman, Nottinghamshire Council for School Sport
Sports Development Officers, local district councils

Other co-operating agencies
County governing bodies of sport
School sports associations
Local Education Authority

Recruiting the children

The children were nominated mainly by their teachers. A few children were invited via their clubs, and children who had been on Nottinghamshire Sports Training Scheme (a Nottinghamshire County Council initiative offering coaching opportunities for young people at a variety of levels) courses were invited to attend the assessment sessions.

Assessment sessions were held at which children were coached and then selected by the head coach and assistant coaches, sometimes with help from teachers.

Contribution from the children
£10.00 for the ten-week programme.

Involving the parents

Parents were invited to attend the assessment sessions, where they were able to talk to representatives from the Sports Training Scheme and local clubs about the opportunities available to the children.

Promoting the scheme

Information about the scheme was sent to PE staff, and some personal contacts were made. The information was passed on to the children by the PE staff. Clubs were contacted by word of mouth via coaches already involved with the scheme. There was lots of discussion with other agencies, such as sports development officers, the education authority and governing bodies.

A press release about the scheme was sent to the local sports council.

Strengths of the scheme

● We are working in a county which already has a flourishing youth sport development plan.

● Many active coaches are willing to take part in the scheme to enhance their own coach education.

Scope for development

We need to forge stronger links with other Nottinghamshire local authority coaching schemes so that Champion Coaching becomes an integral part of the long-term development plan. We also need to encourage the district councils to take on Champion Coaching delivery and become more involved in its planning.

New links

New links are constantly being formed with schools as other PE staff hear about the scheme from those who are already involved. Some district councils have become actively involved and are keen to support Champion Coaching.

Other comments

There has been some direct contact from parents who have heard about the scheme and are keen for their children to participate. Some have asked why they haven't heard about the scheme through their child's school!

The sport programmes

Sport	No. of sport progs	Sex	Ability level	Age group	No. children nomin'd	No. children selected	No. exit routes	No. new exit routes
Netball	3	Girls	Part/perf	11–14	200	110	6	1
Orienteering	2	Mixed	Part/perf	11–13	60	24	2	–
Rugby Union	4	Boys	Part/perf	11–13	90	90	6	–
Squash	3	Mixed	Part/perf	11–13	40	33	5	–

Sport programmes in basketball, cricket, hockey and table tennis started in September.

The facilities

Sport	Facility	Owned by	Hourly rate
Netball	Courts (school)	Notts Education Authority	£38.00 per session
	Courts (leisure centre)	District Council	£1.75 per crt
	Courts (leisure centre)	District Council	£85.00 for ten weeks
Orienteering	School and study centre	District Council	£25.00 per session
	School and woods	Notts Education Authority/ District Council	£15.00 per session
Rugby Union	Pitches (schools)	Notts Education Authority	Equipment vouchers
	Pitch (club)	Newark RUFC	Free
Squash	Courts (clubs)	Squash clubs	£80.00 for ten weeks

The coaches

Sport	No. head coaches	No. coaches	Scholarship work at proposed equivalent S/NVQ level 2		Scholarship work at proposed equivalent S/NVQ level 3	
			NCF	NGB	NCF	NGB
Netball	3	7	4	4	3	3
Orienteering	2	3	–	–	4	–
Rugby Union	4	8	2	2	6	9
Squash	3	4	4	3	4	3

Plymouth Champion Coaching

1991 1992 1993 1994

Youth Sport Manager: Linda Salisbury

Scheme address:
Leisure Department
St Andrews Court
12 St Andrews Street
Plymouth
PL1 2AH

Telephone: 0752 673337

Local management

The Youth Sport Manager is employed by Plymouth City Council as Schools/Sport Liaison Officer.

Time spent on scheme: approximately 2 days per week.

Management structure

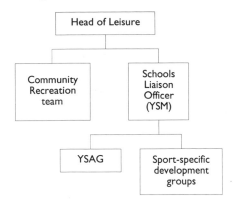

Youth Sport Advisory Group members
YSM
NCF Development Officer
Headteacher
Governor
Cricket Development Officer
Sports Council Regional Officer
Head of PE
PE teacher
Head coach

Recruiting the children

Children were nominated solely by their PE teachers. Assessment sessions were held for the sport programmes in rugby union, orienteering, hockey, table tennis and basketball.

Contribution from the children
£1.00 per two-hour session.

Involving the parents

At the cricket and basketball sport programme sessions, parents began to help coaches.

Promoting the scheme

Schools were contacted with visits to headteachers and PE teachers. Children heard about the scheme through their PE teachers and the local press. Clubs were visited, and coaches contacted through the Plymouth Advisory Sports Council.

There was also general publicity about the scheme in the local press.

Strengths of the scheme

● All secondary schools in Plymouth supported the scheme.

● The scheme has provided the first co-ordinated structure for coaching.

● Coaches have been introduced to PE teachers, and some coaches are now working with teachers in schools.

● Champion Coaching has provided an impetus for the formulation of a coach development strategy.

Champion Coaching in Plymouth is a comprehensive eight-sport scheme. A full programme of press releases advertised it widely in the city area, and, although this is the first year, take-up has been very good. It is significant that the YSM is also the Schools Liaison Officer in the Leisure Department. In all cases the children for the sport programmes have been nominated by schoolteachers with all schools nominating participants.

Tennis has been particularly popular, possibly because courts are available at reasonable cost. The girls' football uses the YMCA grounds; Plymouth is the only scheme which has reported this particular link in the network.

In cricket and basketball some parents have become involved with coaching since their children joined the scheme, and it is hoped that they will take up qualifications in due course.

Strengths of the scheme

● The fact that the scheme is based in education means that we had easy access to teachers and pupils.

● The full-time sports development officer allowed sports to be delegated to specific officers.

● The unit had a history of providing a quality service for coaches.

Scope for development

The scheme could be expanded by the addition of more sports, more programmes in each sport, and by additional funding.

New links

A new link to a badminton club was identified during the Young People in Sport Workshop. We also strengthened links with parents and Leisure Services.

Other comments

Setting up the scheme has been hard work, but has given real job satisfaction. Being based in education has, I believe, given Sandwell a head start in developing Champion Coaching.

The sport programmes

Sport	No. of sport progs	Sex	Ability level	Age group	No. children nomin'd	No. children selected	No. exit routes	No. new exit routes
Basketball	1	Boys	Perf	14	42	20	1	–
Cricket	1	Boys	Perf	12–14	40*	40*	1	–
Netball	3	Girls	Perf	11–14	110	90	5	1
Soccer	2	Girls	Part	12–14	30*	30*	1	1
Volleyball	2	Mixed	Perf	14	30	24	1	1

* Estimates — nomination in progress.

The facilities

Sport	Facility	Owned by	Hourly rate
Basketball	Sports hall (sports centre)	Education	Free
Cricket	Grounds (club)	Club	Free
Netball	Sports halls (schools)	Education	Free
Soccer	Fields (school)	Education	Free
Volleyball	Sports hall (sports centre)	Leisure Services	£9.00

The coaches

Sport	No. head coaches	No. coaches	Scholarship work at proposed equivalent S/NVQ level 2		Scholarship work at proposed equivalent S/NVQ level 3	
			NCF	NGB	NCF	NGB
Basketball	1	6	3	3	2	2
Cricket	2	6	6	6	–	–
Netball	3	6	3	3	6	6
Soccer	2	4	3	3	1	1
Volleyball	2	5	2	2	6	6

Southend Champion Coaching

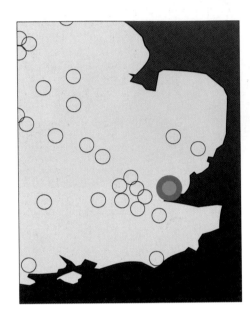

The Southend sport programmes start in September, but preliminary work with schools and clubs is well advanced as More Recipes for Action *goes to press. In this town the local clubs have warmed to the Champion Coaching concept very well, and are playing a significant part in producing coaches and nominating children.*

Facilities are all at local schools, and have been paid for through the medium of the Evans equipment vouchers. These vouchers are available to the schemes at a discount, and their use here will save more than a thousand pounds.

Youth Sport Manager:
Roger Glithero

Scheme address:
Sports Development Team
Civic Centre
Southend on Sea
Essex
SS2 6ER

Telephone: 0702 355612

Local management

The Youth Sport Manager is employed by Southend on Sea Borough Council as Sports Development Officer.

Time spent on scheme: 2 days per week.

Management structure

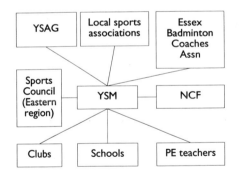

Youth Sport Advisory Group members
YSM
Secretary, Essex Badminton Coaches Association
PE teacher
PE Adviser
Sports Council Regional Officer
NCF Development Officer
Assistant Sports Development Officer, Southend Borough Council

Other co-operating agencies
Essex County Cricket Club
Local coaches association

Recruiting the children

Children were nominated by their PE teachers in all cases. The nominees then attended an assessment session where selection took place.

Contribution from the children
£1.00 per session, with subsidies where need can be proven.

Promoting the scheme

Contact was made with schools at workshops and by attending meetings for heads of PE. PE teachers told the children about the scheme. Workshops were also held for clubs. Coaches were contacted through local coaching associations and personal visits.

Strengths of the scheme

● The enthusiastic and well-qualified coaches, who offered potential help in the exit routes.

● The use of school sports facilities offers a good role model for the community use of schools, and allows other agencies to create greater links.

Scope for development

We could expand by increasing the number of venues for each sport. We will also survey other sports to judge their willingness to take part in the scheme.

Other comments

Owing to Southend's poor facility base, we have had to utilise school sports halls. This has led to teachers becoming involved as coaches in three of the four sports, a fact that will benefit both Champion Coaching and their PE lessons.

Other coaches have been taking advantage of the scholarship scheme — such as cricket coach Roy Gibson, who can start to add qualifications to his years of experience.

There seems to have been a shift in emphasis — from 'what can I get out of this' to 'what can I do to help young people in their sport'.

Photo from London Playing Fields Society

The sport programmes

Sport	No. of sport progs	Sex	Ability level	Age group	No. children nomin'd	No. children selected	No. exit routes	No. new exit routes
Badminton	1	Mixed	Perf	11–14	‡	‡	3	–
Cricket	1	Mixed	Perf	11–14	‡	‡	4	1
Table tennis	1	Mixed	Perf	11–14	‡	‡	2	–
Volleyball	1	Mixed	Perf	11–14	‡	‡	2	1

All sport programmes started in September.

The facilities

Sport	Facility	Owned by	Rate
Badminton	Sports hall (school)	Cecil Jones School	£400 voucher
Cricket	Sports hall (school)	Eastwood School	£400 voucher
Table tennis	Sports hal (school)	Westcliff School for Boys	£400 voucher
Volleyball	Sports hall (school)	Thorpe Bay School	£400 voucher

The coaches

Sport	No. head coaches	No. coaches	Scholarship work at proposed equivalent S/NVQ level 2		Scholarship work at proposed equivalent S/NVQ level 3	
			NCF	NGB	NCF	NGB
Badminton	1	2	–	–	2	2
Cricket	1	2	1	1	2	2
Table tennis	1	1	–	–	2	2
Volleyball	1	2	–	–	3	2

Strengths of the scheme

- Our Sports Development section has much experience of organising coaching courses for children, so we could concentrate our efforts on developing exit routes.

- The sport-specific groups have allowed individuals to make local decisions about their sports.

- Teachers, local coaches and representatives from regional governing bodies and the Youth Service have shared their knowledge and expertise.

- The head coaches were not only committed to the sport programmes themselves, but also to the creation of local exit routes.

Scope for development

In years two and three more sports programmes will take place in different areas of the district. In addition, rugby union, badminton and basketball will be introduced in 1994/5. There is scope for the sport-specific groups to expand their role to take on the co-ordination of sports development and coach education at all levels and for all age groups.

The sport programmes

Sport	No. of sport progs	Sex	Ability level	Age group	No. children nomin'd	No. children selected	No. exit routes	No. new exit routes
Cricket	4	Mainly boys	Part/perf	11–14	94	94	5	–
Hockey	3	Mixed	Perf	12–14	120	120	3	1
Netball*	2	Girls	Perf	11–14	80	80	2	2
Rugby League*	2	Mixes	Perf	11–14	60	60	4	–
Soccer	2	Girls	Part/perf	11–13	30	30	2	1
Table tennis*	2	Mixed	Part/perf	11–14	30	30	2	1
Volleyball*	2	Mixed	Part	11–13	30	30	2	1

*Sport programmes started in September.

The facilities

Sport	Facility	Owned by	Hourly rate
Cricket	Pitch	Wakefield District Met. Council	Free
	Pitch (club)	Ackworth Cricket Club	Free
	Pitch and sports hall	Wakefield DMC Education Dept	£7.50
	Pitch (club)	Leased to club by Wakefield DMC	£5.00
Hockey	Astroturf pitch (club)	Wakefield Hockey Club	Free
	Astroturf pitch (dual use)	Wakefield District Met. Council	£18.50
Netball	Sports hall (school)	Wakefield DMC Education Dept	£18.00
Rugby League	Pitch (school)	Wakefield DMC Education Dept	Free
Soccer	Pitch (dual use)	Wakefield District Met. Council	Free
Table tennis	Sports centre (dual use)	Wakefield District Met. Council	Free
Volleyball	Sports hall (school)	Wakefield DMC Education Dept	£18.00
	Sports centre	Wakefield District Met. Council	£20.00

The coaches

Sport	No. head coaches	No. coaches	Scholarship work at proposed equivalent S/NVQ level 2 NCF	Scholarship work at proposed equivalent S/NVQ level 2 NGB	Scholarship work at proposed equivalent S/NVQ level 3 NCF	Scholarship work at proposed equivalent S/NVQ level 3 NGB
Cricket	4	8	4	4	8	3
Hockey	3	9	2	1	4	3
Netball	2	6	2	4	4	4
Rugby League	2	4	2	–	4	3
Soccer	2	3	3	2	2	2
Table tennis	2	4	2	1	4	2
Volleyball	1	5	1	1	4	3

New links

Many links with some schools and regional governing bodies already existed. These have been strengthened, and new links formed. Identifying and working with coaches has formed good links with local clubs, which in turn have been used as exit routes. The scholarship system has increased coaches' awareness of the materials and courses offered by the NCF.

Other comments

As a result of our soccer programme an under-14 girls team was formed to participate in the 1993 Yorkshire Cup Competition. Taking part was a great experience for players and coaches alike, and gave the girls the chance to put into practice their new skills and appreciation of fair play.

Our hockey programme exemplifies the ethos of Champion Coaching, with strong partnerships between the sub-group and sports development staff, teachers, clubs, etc.

Windsor and Maidenhead Champion Coaching

1991   1992   1993   1994

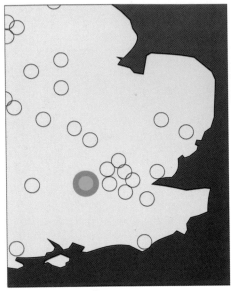

Windsor and Maidenhead was one of the founding schemes in 1991. Only netball and rugby are survivors of the first year, and they are now joined by basketball, hockey, orienteering and swimming.

Orienteering is very much a newcomer to Champion Coaching, but the experience of this scheme suggests that it is a worthwhile one. The enthusiasm generated by its inclusion has played a big part in forming a new junior orienteering club in the area.

Windsor and Maidenhead also makes a feature of including youngsters with a disability in its scheme, and is working with the BSAD to encourage this.

Youth Sport Manager: Lynne Ellis

Scheme address:
Royal Borough of Windsor and
  Maidenhead
Aston House
York Road
Maidenhead
SL6 1PS

Telephone: 0628 796248

Local management

The Youth Sport Manager is employed by the Royal Borough of Windsor and Maidenhead as Sports Partnership Officer.

Management structure

```
        ┌──────────┐
        │   YSM    │
        └──────────┘
             │
   ┌─────────────────────────┐
   │ Sport-specific sub-groups │
   └─────────────────────────┘
```

Youth Sport Advisory Group
There is no YSAG in the Windsor and Maidenhead scheme. Instead, the scheme was managed by sport-specific sub-groups comprising representatives from schools, clubs and governing bodies.

Other co-operating agencies
British Sports Association for the
  Disabled (swimming programme)

Recruiting the children

Schools were requested to nominate up to ten children each to attend a development/promising players' day. Head coaches made selections on those days.

Contribution from the children
£2.00 per session (swimming £2.50 per session).

Promoting the scheme

Headteachers and PE teachers of participating schools were contacted personally. They informed the children about the scheme and made the nominations. Other agencies (clubs, coaches, etc.) were contacted via media coverage and personal contacts.

Articles about the scheme appeared in the local press, and there was publicity in local schools.

Strengths of the scheme

- Links with existing clubs meant that the clubs had a base on which to build their junior programmes.

- Champion Coaching fitted within our existing development plan structure and enhanced provision for this specific sector.

- The use of PE teachers as coaches gave us better access to facilities and children.

- The way in which Champion Coaching has been presented this time round — coaches and young people feel that they are involved in something special.

Scope for development

Additional sports may be added to the scheme, and additional programmes may be added in the existing sports. The funding of these additional opportunities would have to be closely monitored.

Other comments

The Windsor and Maidenhead scheme has been successfully incorporated into the development plan adopted by the Borough. The sports included were selected because a development plan has or was about to be produced, and a working sub-group for each sport met to discuss how Champion Coaching could fit into the existing structure.

The sport programmes

Sport	No. of sport progs	Sex	Ability level	Age group	No. children nomin'd	No. children selected	No. exit routes	No. new exit routes
Basketball*	‡	Mixed	Part/perf	11–13	‡	‡	‡	‡
Hockey*	‡	Mixed	Part/perf	11–13	‡	‡	‡	‡
Netball	1	Girls	Perf	11–13	110	45	4	1
Orienteering	1	Mixed	Part/perf	11–14	90	24	2	1
Rugby Union	2	Boys	Perf	11–13	60	40	2	–
Swimming	1	Mixed	Part/perf	11–14	16	16	3	–

*Sport programmes started in September.

The facilities

Sport	Facility	Owned by	Hourly rate
Netball	Courts (school)	Berkshire Education Authority	£10.00
Orienteering	Various areas	–	Free
Rugby Union	Field (school)	Berkshire Education Authority	£100 voucher
	Field (club)	Maidenhead Rugby Club	Free
Swimming	Pool (private)	Magnet Leisure	Free

The coaches

Sport	No. head coaches	No. coaches	Scholarship work at proposed equivalent S/NVQ level 2		Scholarship work at proposed equivalent S/NVQ level 3	
			NCF	NGB	NCF	NGB
Netball	1	3	3	2	1	1
Orienteering	1	2	–	–	2	–
Rugby Union	2	‡	‡	‡	‡	‡
Swimming	1	3	2	3	2	–

Promotion

One of the advantages of being in a relatively small town is that communication is direct. Adam sees to it that the local papers are kept informed of everything that is going on. During my visit to watch a basketball session at a local school hall, I met a photographer who keeps the local paper 'fed', and I was very impressed when Adam told me that a video was being made to record Champion Coaching 1993 which would be used to promote the 1994 programmes. I mentally winced at the possible cost of this, but my fears were allayed when he revealed that the cameraman was a member of his Youth Sport Advisory Group who just happened to be a passionate video buff and a head of PE!

Adam handles all the marketing and publicity in connection with the scheme, as well as dealing with the nuts and bolts of collecting the cash and booking the halls. He has coaching qualifications in badminton, football and tennis, but it comes as no surprise to find that there simply isn't time to use them.

In case it looks as if youth sport in Ipswich is rather a one-man show, I should again mention the Youth Sport Advisory Group, which Adam refers to quite frequently. This has a very good 'mix' in its make-up, so that as well as giving sound advice and practical help, it is able to prevent wasteful duplication or overlap of provision.

Coaches and coach education

No matter how good the facilities, the success of any youth sport programme is dependent on people, and a supply of keen coaches is as important as anything. You have to press Adam quite hard before he reveals any problems at all with the Ipswich set-up, but I sensed that recruiting enough high-grade coaches could be one of the hardest parts of his job. In common with the pattern throughout the country, ninety percent of his coaches and assistants are volunteers. "There is a falling off in voluntary help", he remarked. "It's often difficult to identify new volunteers. We try to make it attractive by taking all the organising away, so all they have to do is to coach."

Sport 2000 runs coach education programmes in spring and autumn which are based around NCF material, as well as governing-body award courses where appropriate.

All the Ipswich coaches come from around the town, and all are paid a fee — made possible by grant aid from the Eastern Region Sports Council. They are also eligible for the scholarships offered by the NCF through Champion Coaching. These have proved very popular, with eight out of nine coaches taking them up; the ninth is professionally fully qualified anyway.

Adam reports no resentment of paid coaches by the unpaid helpers. It seems that it is now widely accepted that in coaching at least, the labourer is worthy of his or her hire. Probably that is one of the significant social changes within sport in the last couple of decades.

The kids and the cost

All the children were nominated by teachers, and were drawn from Ipswich's eight secondary schools. The teachers' brief was to choose pupils who would be really committed, and fortunately the scheme was able to accept all who were put forward. The cost to each child is one pound per session, and all the courses are for ten weeks, with a session each week. Adam commented that this relatively low charge was made possible because Champion Coaching and Sport 2000 qualified for special community education rates at all the school facilities which were being used. No special transport is laid on; the children use public transport as well as resorting to the common trick of coaxing mum or dad to do some ferrying.

Next year

Steady growth is the plan for 1994. Additional sports to be introduced are boys' basketball, girls' soccer and badminton.

Summing up Ipswich

A small but energetic scheme which maximises its natural advantages. It has a great overall feeling of being very positively managed.

Leafy lanes and synchro swimming awaited our observer in Dorset.

STEP into Champion Coaching

The YSM at STEP, Jane Knowles (centre), encouraging the synchronised swimming group

Deepest Dorset

In spite of having a detailed map, I still managed to make a couple of tours of the ancient town of Wimborne Minster before finding the council offices which house STEP. As these offices are at the top of a leafy lane in a setting apparently designed by the English Tourist Board, my confusion is perhaps excusable.

Jane Knowles, the area YSM, operates from here, co-ordinating youth sport activity over a large area which includes hundreds of square miles of rural Dorset as well as the big coastal towns of Bournemouth, Christchurch and Poole. Although STEP was established before Champion Coaching burst onto the school-age sport scene, Jane saw the opportunities right at the beginning and is a veteran of the first year of the programme. "Champion Coaching phase two is different from phase one", she admitted early on in my day in Dorset. "This time we have to do much more; it takes up lots of time and it's sometimes hard to fit these extra things into the annual work programme."

Jane is very positive about the effectiveness of Champion Coaching. She sees it bridging the sometimes tricky gap between school sport and club sport in a way that has not previously been achieved. "And it can be a huge gap! Champion Coaching is an important stepping stone — it prepares the children in a way that is hard for them to achieve on their own. "Youngsters tend to be afraid of failure if they move on into many of the clubs. This way they are properly prepared for it." Pressed further on the important subject of exit routes, she went on: "Champion Coaching hasn't created many new routes, but it has really strengthened the existing ones."

Jane sees that Champion Coaching has expanded a strictly local scheme into a national one. This helps everyone to 'think a bit bigger', and shows up as added confidence to expand their sporting horizons. She has particularly high hopes of this being successful in cricket, where Champion Coaching participants will be looked upon as potential recruits for junior county squads.

Explaining STEP

STEP stands for 'Sports Training towards Excellence and Performance'. It is a development project which operates in South East Dorset, and is funded jointly by the Sports Council (SW), East Dorset District Council, the Borough Councils of Bournemouth, Christchurch and Poole, and BP Exploration plc. The initial grants were guaranteed for three years, and this term expires in April 1994. The basic idea is to improve the sports performance of committed men, women and children; to co-ordinate effort by strengthening partnerships; to improve the opportunities available for talent, and to improve coach education — all of which fits the Champion Coaching concept perfectly as far as children are concerned.

Coaches

In most of the sports there were enough coaches, but STEP had also taken an initiative in recruiting trainee coaches from the general public. An interesting point that emerged is that more thought needs to be given to exit routes for coaches. It seems that there is a risk of people being encouraged to enter coaching at the first level, who then drift away again because they don't easily fit into governing-body structures.

Jane made one sharp comment about some of the coaches she has met: "Qualifications don't mean that people are able to coach."

The main complaint from coaches is of paperwork overload. We've heard that one before!

Schools

In Dorset, co-operation with schools in general and the PE profession in particular is a recurring theme: teachers are involved at every level, starting with nominating the children for the sport programmes. The majority of the coaches are also teachers, and school facilities play a vital part in the operation. Jane is all for this: she feels that it gives the parents a degree of confidence in the programme which may otherwise be lacking. She acknowledges that Dorset is fortunate — unlike many other areas, teachers here have generally maintained their traditional involvement with after-school activity.

What next?

Jane's biggest hope is for athletics to come under the Champion Coaching umbrella. STEP already operates a Junior School Athletics League, and she believes that this would be greatly strengthened by Champion Coaching participation. At present the choice is between school coaches and club coaches, and the difference can be extreme. She feels that an extra level of 'instruction coaches' would make a big difference in bridging that big exit-route gap.

Jane Knowles

Jane is a graduate of the Chelsea School of Human Movement at Eastbourne. She plays hockey with enthusiasm and spends some of her leisure time learning to persuade horses to jump over obstacles. As Project Manager of STEP, which involves several local authorities and governing bodies as well as outside sponsors, she has to use her obvious diplomatic and managerial skills to the full.

Born in Wales, Jane has worked in Devon and South Wales before coming to Dorset in 1990. Understandably, she loves the county, but I sense that if a similarly challenging job were to become vacant in Wales, the ties of her homeland could well draw her into applying for it.

Publicity and problems

I'm sure that Jane won't mind if I say that she seems to be a great realist. She is doing a big job in a big area, and has a tight budget to do it with. Her publicity cash has to be spent very carefully, so naturally she tries to get as much free coverage as possible from the local press. "I've built up good relations with them, but it's still rather hit and miss. It's a constant effort; I find that you can get so bound up in organising something that you forget to tell people what is going on." I guess we all know that feeling!

There is a problem with some adult clubs which are reluctant to admit junior members. Jane often comes up against this, and if direct appeals are unsuccessful, she attempts to get around it by encouraging the establishment of school clubs and developing junior leagues. This sounds easy when you say it quickly, but the route has not been without its disappointments! Table tennis is a particular problem in this respect, but Jane reckons she'll get there in the end.

Another small problem concerns the age-range covered by Champion Coaching. There are sometimes huge differences between the abilities of eleven-year-olds and thirteen-year-olds, and coaching them together calls for special attention and skill.

Practicalities

Dorset is still a relatively prosperous area, and even though the distances involved can be considerable, there has not been much difficulty in persuading parents to ferry the kids to the sport programmes. The synchro-swimming squad which I visited had been assembled from a wide area, but I gathered that attendance was almost always a hundred percent. A little subtle pressure towards attendance is put on the children via the promise that their Champion Coaching T-shirts will be provided free if they don't miss a session.

Chapter 8

Ealing

Pressure for progress

Champion Coaching drives development forward

*Champion Coaching isn't just another spoke in the wheel — it has become a sturdy hub driving development forward (see diagram on page 154). The Blueprint not only assists the organisation of Champion Coaching schemes, but also can be a model for **all** sport for young people within a community.*

That's a big claim! Let's look at how we justify it ...

The main evidence for Champion Coaching's pivotal role is that it retains its character fully while extending responsibility to the communities involved. Initial direct help to manage the schemes is given, but the Blueprint is then applied locally in the way the community wants it. Quality is assured, but progress is encouraged too.

Developing coach education

NCF courses

The National Coaching Foundation Scholarship Scheme has already encouraged hundreds of coaches to acquire formal qualifications for the first time, or to upgrade their existing ones. Applicants are given vouchers for NCF courses which they take in their own time. However, rather than just leaving it at that, many communities have been creative in organising NCF-based courses for their scheme coaches — and their reward has frequently been a take-up of one hundred percent. Some have begun to extend this initiative far beyond the original programmes.

These are just a few examples which show how Champion Coaching has urged coach education forward:

In Plymouth, YSM Linda Salisbury organised the NCF key courses *Coaching Children* and *Motivating your Athlete*. She already had enough coaches from Champion Coaching to justify running the courses, but decided to advertise through schools and the Plymouth Advisory Sports Council to see if any others were interested. They were! Forty-two extra places were taken on the

two courses. These were the first-ever NCF courses to be organised by Linda, and she now has enough interest to run a rolling programme throughout the year.

In Humberside, YSM Mark Nesti thought of another development: his team of SDOs have been trained to present NCF courses. Now Mark has a team of tutors to run courses whenever he needs them, and the SDOs have all gained another stage of professional development.

Ipswich and St Edmundsbury have co-operated: they are close enough together to share courses. The key organisers, Adam Baker, Sharon Ayres and Alison Furlong, regularly meet to discuss the programme, and as a result have increased the numbers on courses while reducing their own administration loads.

National governing body courses

Increasing sport-specific knowledge

Just as the scholarships for the general NCF courses are proving popular, coaches are also taking up extra national governing body (NGB) training as a result of Champion Coaching's scholarship system. Nationwide, this has meant an increase in communication between NGB liaison officers and YSMs, as well as deepening coaches' knowledge. The comment we receive most frequently is that communities, particularly the rural ones, are now able to make their needs known to the right people at the NGBs in a way that never happened before Champion Coaching came on the scene. Once again, co-operation between schemes has been a key feature of the organisation of NGB courses:

In the South West Region the Youth Sport Managers collaborated to decide which NGB courses they needed. Each YSM was then took responsibility for organising one of them.

A hockey course which was arranged for Champion Coaching was also advertised nationally by

Hockey at Grimsby, Humberside

A breather during tennis at St Edmundsbury

the NGB. As a result the course was completely filled and Alan Bywater, the YSM for Bath, used it creatively:

I was aware that I did not know some of the coaches attending the course, so I went along. Having watched the course and seen the coaches in action I approached some and asked if they would like to be involved in Champion Coaching. The course gave me a chance to see the way they worked.

Mentoring

A new word in coaching

Mentoring is our word for the process whereby experienced coaches give support and advice to less-experienced colleagues who are climbing up the coaching ladder.

Mentoring is an emerging technique in UK sport, but its development has been accelerated by Champion Coaching. Initially head coaches were encouraged to become mentors, but master coaches have now also become involved. Lynne Goodliffe from hockey and Tony Butlin from rugby union have agreed to 'mentor the mentors'. Both are highly respected in their sport, and their reactions will be used to develop mentoring even further.

Workshops in mentoring have been conducted by experienced NCF tutors in every region, and over 500 head coaches have attended. Feedback from tutors and coaches at these workshops is being used by the NCF to develop a course in mentoring at the key-course level.

NGB workshops

Several national governing bodies have organised training for Champion Coaching coaches. As well as being excellent opportunities for the NGBs to meet coaches, these also give the coaches the chance to exchange ideas and feel part of the extended Champion Coaching family. Some NGBs have seen our workshops as ideal showplaces for their latest developments. Netball and volleyball have been good examples: the most senior staff members of these NGBs used Champion Coaching training days as the occasions to unveil the latest youth coaching and safety developments in these popular sports.

Youth Sport Advisory Groups (YSAGs)

Youth Sport Advisory Groups inevitably include members who are influential at a variety of levels within their respective communities, so it is not surprising to find that they are broadening their activities. This is happening in three main ways:

● Some are speeding the development of youth sport generally by extending their roles to include opportunities for youngsters who are above or below the ability range of Champion Coaching.

● In some of the county-wide schemes, the main YSAG is overseeing a number of other ones — usually one in each district or borough. Kent is a good example of where this is happening, and here the district YSAGs are being used to find out how ready the districts are to run their own Champion Coaching schemes in future.

● Certain YSAGs have involved sport-specific sub-groups in the local organisation of Champion Coaching. Where this has occurred, the sub-groups have been able to integrate the message of Champion Coaching into their other work.

By adapting in these ways, the YSAGs are playing a big part in driving youth sport forward.

Resources

Sport Logs boost communication

The 20,000 *Sport Logs* which were distributed to the Champion Coaching children in 1993 represented a huge opportunity for youth development. The *Sport Logs* contain a form on which coaches are encouraged to make statements that relate to Key Stage 3 of the National Curriculum. Teachers can use these to complete the End of Key Stage statements for each child, and in this way the work of coaches and teachers becomes complementary. As the partnership between teachers and coaches continues, the vital understanding between the two groups will increase.

The *Sport Logs* have proved so popular that many YSMs seized on them as a way of conveying information about other local programmes. And now we are delighted to see that they are beginning to be used for related schemes which are just outside Champion Coaching.

Workshops broaden the schemes

The range of Champion Coaching community workshops is one of its most useful resources. There are workshop programmes for parents, for clubs, and for

headteachers and governors, and all of these are developing in many ways. Tutors are being trained by Champion Coaching to ensure consistent delivery. The impressive group which we have recruited to present the headteacher and governor workshops includes PE Advisers, heads of Leisure Services Departments and Principal Officers. They are all able to influence heads and governors at the highest level.

Champion Coaching's range of resources is comprehensive, and it is pleasing to see that YSMs have used them creatively.

Coaches and children take time to fill in their Sport Logs *after a session in the pool (South-East Dorset).*

Champion Coaching — driving development forward

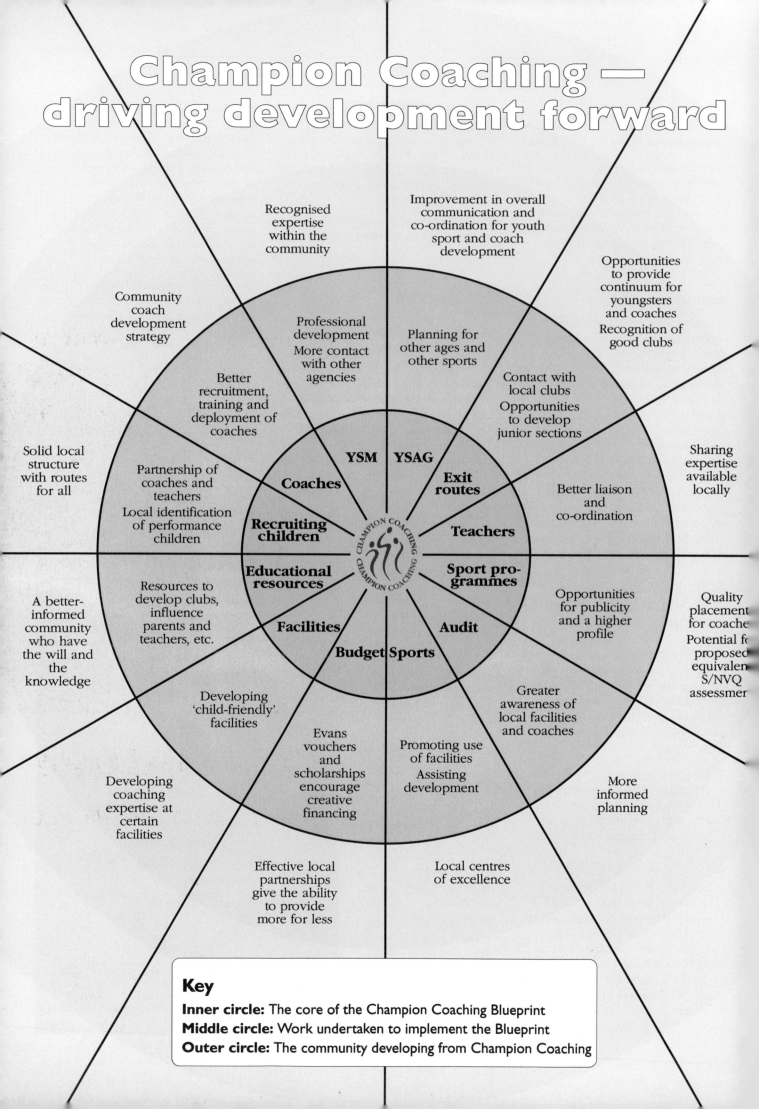

Key

Inner circle: The core of the Champion Coaching Blueprint
Middle circle: Work undertaken to implement the Blueprint
Outer circle: The community developing from Champion Coaching

You Can... Use language in a positive way

Because so much of a teacher's work involves verbal communication, it is really important to use language in an effective way. The positive use of language can help you control your class, give clear and understandable instructions, and will generally help you keep a calm atmosphere in your classroom.

Thinking points

● Teachers often use rhetorical questions, where an answer is not actually required or desired. I often catch myself saying, *Why are you doing that?* when I don't really want to know the child's motivation.

● Focusing on a positive use of language will help you to avoid the natural human tendency to nag at your children when they are not doing as you wish.

● Positive language will encourage your children to follow your instructions and will help you give clear lesson explanations.

Tips, ideas and activities

● Train yourself to use statements of what you *do* want, rather than negative complaints about what you don't. These positive statements will help you maintain a more confident and effective teaching style, and they will encourage your children to work and behave as you wish. Here are a couple of negative and positive examples, to show you what I mean:

   ● Negative: *I can't believe you haven't finished that work yet.*
   ● Positive: *You've got three minutes to finish the next five questions.*
   ● Negative: *Why are you all talking again – haven't I told you to be quiet?*
   ● Positive: *That's great, Ben – you're sitting really quietly. Now let's see who else wants to go to break on time.*

● Develop your own set of positive cues to use with your class. These are statements that anticipate the required behaviour. So rather than saying, *Who can tell me the answer?* (and having lots of children call out), you might instead say, *Put your hand up if you can tell me the answer.*

● Devise a shorthand with the class, whereby more complex instructions are simplified to short (but polite) commands. You might say, *Hands, please!* to get answers to a question, or, *Eyes, please!* rather than, *I want everyone looking this way.* Combine these instructions with non-verbal cues, such as raising your own hand or pointing to your eyes.

● Repetition is a very powerful linguistic technique for helping to create a calm and focused atmosphere. You might:

   ● repeat an instruction several times to ensure that everyone understands what you want (to avoid the familiar complaint, *I don't know what to do.*)
   ● repeat a child's name until he or she looks at you – names are very powerful in pulling children's focus
   ● use the 'broken record technique' to ensure you have the full attention of the class (this technique involves pausing and then repeating the same phrase over again, until you are sure that all the children are listening).

You can ... Make more use of non-verbal communication

Teachers can 'say' a great deal without opening their mouths – some classes will even fall silent when the teacher simply stands and stares. Just as clear and positive language is vital for effective teaching, so we also need to make the best use that we can of non-verbal communication. The less we have to speak to manage the class, the less likely we are to be drawn into nagging.

Thinking points

● If you can calm and control your class without talking, the children will get the strong impression that you are in charge. Effective non-verbal communication appears almost like a form of 'magic', where the teacher manages behaviour and learning, apparently without doing anything at all.

● Non-verbal signals are very useful because you can use them privately to warn an individual, without the rest of the class being aware of the interaction.

● Your body language will signal how confident you feel about your teaching and your classroom management skills. Children are very good at subconsciously reading the signals that teachers send with their bodies.

Tips, ideas and activities

● Your face is a powerful component of your teaching style. Work on your 'deadly stare', combining eye contact and facial expression to tell your class or an individual that you are not impressed. If you can raise a single eyebrow, this is extremely effective. Keep your eyes ranging around the room, constantly checking for any signs of inattention or misbehaviour.

● Consider your overall posture and what this might communicate to the class. Folding your arms or putting your hands on your hips are typical teacher stances to indicate *I'm waiting*. Standing up tall will help the children feel that you are confident and in control.

● Use an egg-timer as a visual indicator of wasted time. Every time you have to pause and wait for the children's attention, turn the timer over so that some sand runs through. When the class returns their attention to you, drop the timer on to its side. At the end of the lesson, you can insist that the class earns back the total amount of time wasted.

● Take care about sending the wrong messages. Avoid defensive postures, such as backing up against a wall or getting trapped into a corner. Try not to drop your head or break eye contact, as these indicate uncertainty.

● Your hands are particularly expressive, so make lots of use of them. You might:
  - push your hands palm downwards to indicate *calm down*
  - click your fingers to get the children's attention
  - give a 'thumbs up' signal to say *well done*.

● Avoid pointing directly at individual children. This can be interpreted as an aggressive signal.

● Adapt your body language and posture to suit formal or informal times in the classroom. For instance, you might lean or sit on a desk to have an informal chat with the children.

You Can... Calm down overexcited children

Inevitably, there will be times in your classroom when you wish to do a particularly exciting activity. Of course, there is absolutely nothing wrong with children getting excited – it is great to see them engaged by and interested in their learning. However, teachers have to constantly be aware of potential safety issues. In addition, where a class loses all sense of self-control, this can mean that a normally calm atmosphere gets destroyed.

Thinking points

● Where a class becomes easily overexcited, it can be tempting for the teacher to avoid engaging in unstructured activities. This is a shame, and it also means that the children do not learn to control their own responses to exciting situations.

● It is important to build your children's sense of self-control and self-awareness. This is all part of teaching them to take responsibility for their own behaviour.

● Excitement is very infectious – one child starts to scream and, before you know it, the whole class has joined in.

Tips, ideas and activities

● Use the allure of exciting activities to encourage the children to manage their own behaviour. Before you begin a fun exercise, discuss your expectations of behaviour with the class. Explain that, if they cannot take the work seriously and stay under control, you will have to stop the activity. When you make this threat, you *must* follow through and carry it out if the children do not behave as you wish.

● Practice makes perfect – accept that, the first few times you try an unusual or exciting activity, the children might get overenthusiastic. Put yourself in their shoes and don't forget how it felt to be a child. As they gain in experience, your children will learn to control themselves.

● If your class is getting overexcited, take a moment to control your own responses before you intervene. If you are annoyed or panicky, you will not be able to deal with the situation in a rational manner. Take deep breaths and think carefully about the most appropriate intervention to make.

● Focus on the way that you use your voice to help you calm the class down. Try a slow, almost hypnotic tone and try to use lots of repetition.

● Where a class is getting completely out of control, use a sharp sound signal to gain the class's attention. This might be a whistle, a horn or some loud hand claps. Whatever you do, don't attempt to shout over the racket – the children are unlikely to be able to hear you.

● Alternatively, try writing a long sentence on the board – this is usually a great way of pulling the class's attention to the teacher. You might write: *I want everyone to calm down by the time I finish writing this sentence, because, if you don't, you will force me to stop this work and do something a bit less interesting and that would be a real shame.*

You Can... **Build teamwork and cooperation**

In an environment where there are a large number of people working together, cooperative attitudes are vital for building positive and harmonious relationships. Use lots of activities that require cooperation to be successful, and build a sense of community within your classroom.

Thinking points

● Children need to learn how to cooperate with others. In the earliest years, children will play alongside each other, but will not understand how to play together.

● A key part of children's development is learning how to work well with their peers. Where a class works together successfully, this helps to promote a positive learning environment.

● Children can be quite unforgiving of those who are 'different' in some way. Insist that your pupils treat everyone and anyone with equal respect.

Tips, ideas and activities

● Physical exercises and drama activities are great for promoting cooperative work. Here are some ideas:

  ● Split the class into small groups and give each group a balloon. The aim is to keep the balloon in the air for as long as possible. Each child must touch the balloon in turn, and they cannot touch it again until everyone else has had a go.

  ● Divide the children into groups of three. Two of the children stand facing each other and the third child stands in the middle. Keeping as straight as possible, the middle child 'falls' backwards on to the first child's hands, and is then pushed forwards on to the other. (This can also work with a small group standing in a circle.)

  ● Stand with the class in a circle and place a chair in the centre. The children must work together as a team to stop you from sitting down in the chair, by going into the circle and using the chair as a prop (for example, a car or a horse). You can only sit down if the chair is not already occupied.

  ● Create an obstacle course, using chairs, tables and whatever else comes to hand. Split the children into pairs. One child closes his or her eyes and the other child guides the partner around the obstacle course by whispering instructions.

● Try some cooperative story writing.

  ● With the children in a circle and ask each child in turn to add one word or sentence to create a whole-class story.

  ● Give each of the children a piece of paper and ask them to write the first line of a story. Each paper is then passed on to the next child, who writes the next line. At this point, the paper is folded over, so that only the second line of the story is visible.

● When organising groups, insist that the children must be willing to work with anyone in the class. If you use mixed-gender groupings right from day one, then the children will see this as the natural way to work.

You Can... # Help your staff achieve a work/life balance

The staff are the backbone of the school – without them it cannot function. Sensible managers put a priority on treating their staff well, taking account of their needs both outside and inside of school, and helping them find a balance between work and home life. Without this balance, it is likely that teachers and other staff will succumb to stress and some will even leave the profession.

Thinking points

● Where staff feel valued and cared for, this can lead to improved staff retention rates and lower levels of sick leave. Overall, achieving a good work/life balance for staff is likely to save, rather than cost, money.

● With the introduction of the Workload Agreement in England, teachers are (or should be) getting help in completing routine non-teaching tasks and being given non-contact time for planning.

● Where teachers are stressed about their own domestic concerns and feel that they do not have time to fit everything in, this will probably lead to a tense and irritable approach in the classroom.

Tips, ideas and activities

● Some headteachers have found really innovative ways of supporting their staff. Why not suggest that your own school try some of these approaches and see their positive impact:
  ● organise dry-cleaning, laundry and ironing services
  ● book a masseur to come into the staffroom and give head massages in break and lunchtimes
  ● offer group discounts on gym membership for staff
  ● offer experienced staff a secondment overseas to give them an interesting new challenge
  ● send staff home at 4.30pm and insist that they take time off to relax
  ● ask the canteen to prepare 'ready meals' for busy teachers to take home in the evenings.

● Encourage staff to act as a team, to cut down on the duplication of work. Ensure that schemes of work are clearly written and easily accessible, so that they can be reused by new teachers over a period of time.

● Get teachers and support staff to play to their talents. For instance, a teaching assistant who can play the piano well might take a class for a singing session, while the teacher spends time on planning and preparation.

● Consider employing a specialist on a part-time basis, to free up non-contact time for teachers, for instance, a sports or music teacher who can deliver these subjects to the children one afternoon a week.

● Find ways of freeing teachers from mundane duties, such as playground supervision and exam invigilation. These are not effective ways to use a professional person's time and energy.

● Ensure that meetings are effectively planned and free from 'waffle'. Have clear objectives, just as you do for lessons.

You Can... **Keep it calm in the corridors**

Although much of your time will be spent in your main classroom base, there will be a number of times during the average week when you wish to move your children from one part of the school to another. Keeping a sense of order in the corridors will help contribute to your overall ethos of calm.

Thinking points

● The way in which your pupils move through the corridors will influence the state of mind in which they arrive at lessons.

● A key time is when the children come into the school after a break. If the move from outside to inside is not well managed, time will inevitably be wasted in settling the children down when they arrive in the classroom.

● Young children are very easily influenced by the behaviour that they see from older pupils in a school. Where there is a culture of 'rough and tumble' around the school building, this can quickly percolate down the year groups.

● In an open-plan school, it is particularly important that the children know how to move calmly and quietly around the building. This is especially so when they are moving during lesson time.

Tips, ideas and activities

● It is a great idea to bring in a one-way system in your corridors, especially if space is tight. You might devise and introduce this new corridor system during lesson time, as part of a mathematics or design and technology project.
  ● In mathematics, you might measure the width and length of the corridors, calculate the number of pupils passing through at certain times of day, construct frequency tables and so on.
  ● In design and technology, you might generate and develop ideas for a new corridor design, coming up with ideas for suitable graphics to indicate the direction in which the pupils should move.

● Work out the 'flashpoints' in the corridors and ensure that these areas are supervised when pupil traffic is heavy. Typical problem areas include the top and bottom of stairs, blind corners and waiting places for access to canteens.

● Think laterally about how you might develop your corridors into more interesting and relaxing spaces. For instance, is there an unused corner where you could site a few chairs and display some interesting artwork? This might develop into a spot where the pupils can sit quietly during a wet playtime.

● Establish clear expectations about the way in which your children will move through the corridor. You might even have a short set of 'corridor rules', which apply right across the school. Consider what sanctions you are going to use if and when these rules are broken. Because the corridors are a communal space, a suitable sanction might take the form of a 'community order', whereby the children have to clean up litter.

● Harness your children's imagination when moving your class from one place in the school to another. For instance, ask them to move in slow motion, or as though they are walking on the moon.

You Can... **Use circle time to promote concentration**

Circles offer a wonderfully democratic format for discussion. Everyone in the class can see everyone else, and there is a sense of equality and community. Circles are also very useful for activities where you want everyone to contribute, as you can move freely in a clockwise or anticlockwise direction around the class. Encourage your children to concentrate fully during circle time, building up gradually to longer periods of listening.

Thinking points

● Although they should not be used solely as a control method, circles are in fact very useful for establishing appropriate behaviour and promoting a sense of calm.

● The format of a circle allows the teacher to see all the children and to work with them within a confined space.

● Circles can be very useful as a 'gathering' position during an active lesson. On the teacher's signal, the children come back to sit in a circle to be given further instructions.

Tips, ideas and activities

● When you first use the format of a circle, be sure to set up some basic ground rules. It is important that everyone in the circle shows respect, and this means listening in silence when others are speaking and always receiving contributions in a positive manner. I would also give the children the option of 'passing', if they do not want to contribute to an activity. This helps the shy children gain in confidence before they start to join in.

● A great physical warm-up for a circle session is a game called 'Pass the clap'. A clap is passed around the circle – to the left by turning to the left and to the right by turning to the right. When you are passed the clap, you can send it on in either direction. Start by passing a single clap – the idea is that the sound is continuous and smooth rather than staccato. Once they get the hang of it, go on to having two claps moving around at the same time.

● When you ask a question that requires a verbal response, it can be useful to have a few moments 'thinking time' before you ask for answers around the circle.

● Warm up your children's brains by asking a simple *What's your favourite?* question as a starter exercise. This might be their favourite: food, subject, book, animal, television programme.

● Sometimes you may want to let volunteers contribute their own ideas in a more random way. It is a good idea to use a 'conch' to ensure that the children don't just talk over each other. The conch is a special-looking item of some sort (for example, a shell or a decorated stick), which gives the holder the right to speak without interruption. When the child has finished speaking, he or she places the conch back in the middle of the circle so that someone else can have a go.

You Can... **Make use of all the senses**

Where the learning appeals to all of your children's senses, they feel a far greater connection with the work. Much of the time at school we ask our children to work with their minds and their intellects. Reconnecting to the instinctive world of the senses can help them feel a stronger relationship with the world around them.

Thinking points

● Sensory responses can be used to help us feel calm, for instance: being surrounded by soft textures, listening to peaceful sounds, such as waves or birds singing, or through the use of aromatherapy and special scents.

● Children who have special needs, such as autism, will benefit a great deal from being given a range of sensory experiences.

● Looking for ways to incorporate the senses into your classroom will make your teaching more interesting and imaginative.

Tips, ideas and activities

● Here are some suggestions for ways to incorporate sensory responses into your teaching.

  ● In a science lesson on materials, blindfold some volunteers and ask them to identify materials through their sense of touch. Talk about how we know, simply by touching, that a material is made of metal. Brainstorm the vocabulary connected to different materials.

  ● Begin a creative writing session by getting the children to smell a range of different scents, such as herbs, coffee, and washing powder. Ask the class to talk about the atmosphere that these smells create, and the type of people and locations that they associate with them.

  ● Bring some fruit and vegetables into your classroom for the children to taste, to start off some work on healthy eating. (Remember to check for allergies and dietary requirements first.)

● The photocopiable on page 58 – 'Me and my senses' is designed to get your children thinking about their own sensory responses and the emotional reactions that their senses evoke.

● The first activity on the photocopiable page asks the children to look at some pictures. Ask the children to talk with a partner about how the items in the pictures might be used to help them stay calm. Then they should write in their ideas below the pictures. Depending on the age of your children, you might like to discuss this worksheet as a whole class first, writing up some key words on the board. Here are some suggestions for ideas that your children might give.

  ● Cat: stroke it to feel calm; sit still with the cat on lap.
  ● Cushion: hit it to release tension; sit down on it to feel relaxed and comfortable.
  ● Flower: smell it to enjoy the scent; pick petals to make perfume; give it to friends to make them happy.
  ● Sunshine: sitting in the sun to feel warm and relaxed.

● Next, the children are asked to find some images, textures, scents, sounds and tastes that make them feel good. They might draw these, cut out pictures from magazines or even stick items on to the worksheet (for instance, a sprig of rosemary for scent, a sweet for taste and so on).

'Reading' people

You are very late for work and you are about to miss your bus.	You are very shy.
You have a very high temperature.	You have a thumping headache, and you are in a very bad mood.
You are desperate to go to the toilet.	You think someone is following you.

Name: _____   Date: _____

What makes me cross?

Talk with a partner about all the different things that make you feel cross.
Write some ideas down here.

_____

_____

Now look at the pictures below and work out what is going on.

With your partner, answer these questions:

● How are the children in these pictures feeling?
● What is making them feel like this?
● What can they do to feel better?
● What can other people do to make them feel better?
● What would you do if you were in this situation?

Choose one picture and write a short story about what is happening. Try
to include lots of emotion and interesting language. Plan your story here.

_____

_____

_____

Name: _____ Date: _____

Calm words, angry words

Here is a list of words. Some of these words are used to talk about calm feelings and some of these words are used to talk about angry feelings. Read the list and use a dictionary to look up any words that you do not understand.

Calm   Soft   Livid   Mad   Cool   Upset
Cross   Still   Angry   Snappy   Quiet   Furious
Wound up   Peaceful   Gentle   Irritated Serene   Relaxed

Write the words out on the table below. Put the calm words on one side and the angry words on the other side.

Calm words	Angry words

Now write sentences that include some of these words. Here is an example: The boy lay on the beautiful beach, feeling calm and relaxed as he listened to the sea.

_____

_____

_____